Reviews of *The Quantum Menagerie*

An informative and accessible introduction to the bizarre world of quantum mechanics and the different interpretations of the mathematical formalism. Stone's clear, organised style makes difficult concepts seem simple.
Ian Stewart, Fellow of the Royal Society, Emeritus Professor of Mathematics, University of Warwick.

A wonderful resource for anyone who wants to teach themselves quantum mechanics for real. You will learn the necessary math, but the real emphasis is on conceptual understanding – Stone's book explains why things work a certain way, rather than just asserting that they do. It dispels the mystery from an intimidating subject.
Sean Carroll Research Professor, Caltech, author of *Something Deeply Hidden.*

The Quantum Menagerie is a wonderfully clear introduction to the notoriously demanding subject of quantum mechanics. Uniquely it blends the history of the field, including the work of Planck, Einstein, and others, with a splendidly lucid, step-by-step approach to the maths behind its key findings. James Stone explains milestone results that might appear abstract at first glance, such as Bell's inequality, in a delightfully visual manner. Recommended to anyone who would like to understand the formalism of quantum mechanics and needs the guidance of a seasoned explorer.
Professor Paul Halpern, author of *The Quantum Labyrinth.*

In this lively and entertaining book, James Stone traces the development of Quantum Mechanics, explaining how its salient features were born from pure guesswork, or, in some cases, sheer desperation, as scientists faced observations that refused to fit into the framework of classical physics. Dr Stone describes the problematic aspects of Quantum Mechanics, and the failed attempts to fix them, which in some cases led to the experimental confirmation of some of Quantum Mechanics more mind-blowing predictions. This book is written at a level suitable for beginning undergraduates.
Richard Fitzpatrick, Professor of Physics, University of Texas at Austin.

The Quantum Menagerie is a terrific tour of the mathematical underpinnings of quantum mechanics. It provides a gentle introduction to unfamiliar mathematical concepts and their application to quantum physics. This book has been particularly useful to biology students within the Quantum Biology Doctoral Training Centre, most of whom have little training in mathematics. Professor Johnjoe McFadden, Director, Leverhulme Quantum Biology Doctoral Training Centre, University of Surrey, UK.

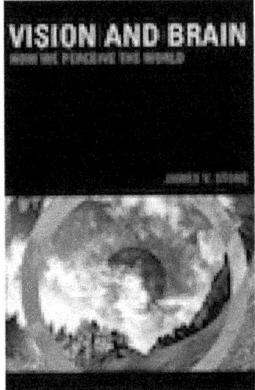
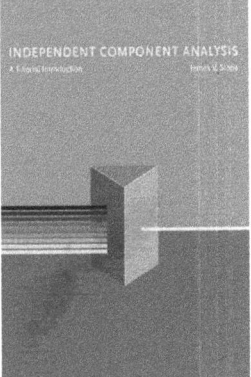

James V Stone is an Honorary Associate Professor at the University of Sheffield, UK.

The Quantum Menagerie

A Tutorial Introduction to the
Mathematics of Quantum Mechanics

James V Stone

Title: The Quantum Menagerie
A Tutorial Introduction to the Mathematics of
Quantum Mechanics
Author: James V Stone

©2020 Sebtel Press

First Edition, 2020.
Typeset in LaTeX$\partial 2_\varepsilon$.
First printing.

ISBN: 9781916279131

Cover background by Criminalatt.

For my brother, Bob.

There, in our long goodbye
We spoke those gentle words
That don't come easy to London boys
But whose truth we had always known
Through all these best years of our lives.

There is a theory which states that if ever anyone discovers exactly what the Universe is for and why it is here, it will instantly disappear and be replaced by something even more bizarre and inexplicable.

There is another theory which states that this has already happened.

Douglas Adams

Contents

Preface

All the paths up Mount Quantum lead to the summit, but whereas some paths are short and steep, requiring considerable expertise, others are long and shallow, requiring little expertise but some fortitude. This book represents a long and shallow path that will guide the diligent traveller to a summit from which the views are without equal.

Books on quantum mechanics come in two basic formats: popular science books and textbooks. In contrast, this book represents a middle way between these formats, by combining the informal approach of popular science books with the mathematical rigour of introductory textbooks. Accordingly, the material in this book concentrates on the basics, and covers more advanced material only when these basics have been fully explored. To aid understanding, over 50 diagrams are included. Additional support is provided in the glossary, tutorial appendices, and annotated list of further readings.

Unlike traditional texts, topics are presented in order of increasing mathematical sophistication, rather than historical order. This allows ideas covered in one chapter to act as a mathematical foundation for ideas presented in subsequent chapters. To compensate for this admittedly unusual approach, a brief history of quantum mechanics is provided in Chapter 8, and a guide to more detailed histories is given in the annotated Further Reading section.

Who Should Read This Book?

The material in this book should be accessible to anyone with an understanding of basic calculus. The tutorial style adopted ensures that readers who are prepared to put in the effort will be rewarded with a solid grasp of the fundamentals of quantum mechanics. Upon reaching the end of this book, readers should be equipped to tackle more conventional textbooks on quantum mechanics.

Additionally, having read this book, attentive readers will not only understand the essential elements of quantum mechanics but also be able to appreciate this well-known physics joke:

Heisenberg is driving Schrödinger home when they get pulled over. The policeman says to Heisenberg, "Do you know how fast you were going?" "I have no idea," says Heisenberg, "but I do know exactly where we are." The policeman then informs Heisenberg, "You were speeding at precisely 66.26 miles per hour." Heisenberg throws his hands up in despair, "Great! Now we are completely lost." Worried by this bizarre reaction, the policeman orders him to open the boot (trunk) and wanders over to take a look. "Hey, do you know there's a dead cat back here?" "We do now!" says Schrödinger, shaking his head.

Like most physics jokes, this is not very funny, but it is clever. And don't worry, there are no more jokes.

Acknowledgements

Thanks to John de Pledge, Royston Sellman and Stephen Snow, who scratched their collective heads over the problems of quantum mechanics, and for the beer we had to drink to make sense of those problems. Thanks to Philip Nelson for recommending material on Planck, and to Eliahu Cohen for recommending papers on information theory and the double-slit experiment. Thanks to Tikz writers for allowing me to adapt Figures 3.1 (Izaak Neutelings) and 3.2 (Cyril Langlois). Thanks to Teleri Stone for subduing our cat Gypsy for long enough to take the photographs on the cover, and to Sebastian Stone for the picture of Gypsy's 'thousand yard stare' in Figure 1.4. Thanks to Tomas Bzdusek for creating Figure 5.6. Finally, thanks to Johan De Schrijver, Nikki Hunkin, and Amy Skelt for invaluable feedback, and to Alice Yew for proofreading both the words and the physics.

Book Corrections

Please email corrections to j.v.stone@sheffield.ac.uk.

A list of corrections can be found at:

`http://jim-stone.staff.shef.ac.uk/QM/Corrections.shtml`

Nature is stranger and more wonderful than we had once thought or could possibly have imagined.
Mermin ND, 1981.

Chapter 1

What Is Quantum Mechanics?

> *It is my task to convince you not to turn away because you*
> *don't understand quantum mechanics. You see, my physics*
> *students don't understand it either. That is because I don't*
> *understand it. Nobody does.*
> Feynman R, 1966.

1.1. Introduction

Quantum mechanics explains how the physical world works. And because all modern technology, from computer chips to brain scanners, depends on physics, quantum mechanics explains how all modern technology works. More importantly, life in all its forms most beautiful, from bacteria to cats, also depends on quantum mechanics. In short, everything depends on quantum mechanics.

Quantum mechanics is traditionally associated with physics at extremely small scales. Even though it seems entirely plausible that the motion of an electron depends on quantum mechanics, it is less obvious how quantum mechanics could be involved in a system as massive as a star (or a cat). Similarly, it is not obvious how quantum mechanics could play a role in the Darwin–Wallace theory of natural selection, which acts on whole organisms. But quantum mechanics is not only involved in stars, cats and evolution, it is essential for their existence.

The vast theoretical edifice of quantum mechanics depends on a few remarkably counter-intuitive facts: 1) particles such as photons, electrons, atoms and even whole molecules can behave like waves; 2) light waves can behave like particles; and 3) the energy of both waves and particles exists only as small packets, or *quanta*.

Unfortunately, the mathematics of quantum mechanics is formidable. Despite this, the strange physical phenomena it explains are relatively straightforward to describe. Accordingly, a summary of the better-known quantum mechanical phenomena is given here, and mathematical accounts follow in later chapters.

1.2. Spooky Action at a Distance

The phrase *spukhafte Fernwirkung* (spooky action at a distance), or *non-locality*, was coined by Einstein as a direct attack on quantum mechanics. It refers to the seemingly impossible prediction that if two particles are *entangled* then measuring the state of one particle *instantly* affects the state of the other particle, even if they are separated by astronomical distances. The idea of entanglement was proposed in a famous 1935 paper by Einstein, Podolsky and Rosen, which is now known as the *EPR paper*[10].

For example, the state of a photon can be given in terms of its polarisation angle β, which is explained in detail in Chapter 3. For our purposes, these details matter less than the fact that the polarisation angles of two photons can be entangled. Consequently, when the polarisation angle of one photon is measured as β, its entangled partner instantly adopts a particular polarisation angle, irrespective of the distance between the photons. For simplicity, we assume that both photons adopt the same polarisation angle here. No-one understands how this *non-local* interaction could possibly be true; it just is. However, one way to visualise how it could happen is to assume that the entangled photons are projections of a single composite photon in a high-dimensional space, as depicted in Figure 1.1. This figure is not intended as an explanation, but simply as a geometric sketch of the counter-intuitive facts.

A common counter-argument to the whole idea of entanglement is that the two particles simply had the same state when they were created, so naturally their states will be identical whenever they are measured. For example, if the state of one member of a pair was measured as $\beta = 45°$, then the other member of that pair would also have a measured state of 45°. And even though different pairs have different states, both

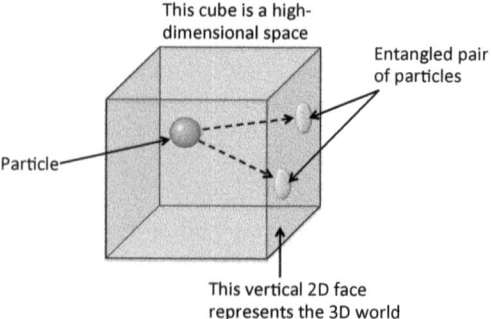

Figure 1.1: Two entangled particles could be regarded as projections of a single particle embedded in a high-dimensional space.

members of each pair have the same state. When considered over many pairs, measuring the state of each member of each pair would yield a perfect correlation. This argument is known as the *hidden-variable hypothesis*, because if two photons had the same state from the outset then the correlation between their states must have been determined by the value of a hidden variable that both photons had all along.

However, there is an ingenious counter-counter-argument, in the form of John Bell's 1964 paper[4], which is both subtle and quite deadly. In essence, that paper defined *Bell's inequality*, which shows how to design an experiment that can rule out any hidden-variable hypothesis. In such an experiment, if two photons had the same state from the outset, and if the states of many such photon pairs are measured as above, then Bell's theorem implies that the correlation between states must be less than a certain critical value.

The first test of Bell's inequality was carried out using photons by John Clauser and Stuart Freedman in 1972, and subsequent experiments have been done using electrons. For both photons and electrons, the correlation between particle states was larger than predicted by Bell's inequality. In other words, the correlation between particle states observed in experiments could not possibly be obtained if those states were set at the time the particles were created.

A disturbing corollary of such experiments is that before the state of a particle is measured, it is not only unknown but also *undefined*. To discuss a particle in an undefined state is difficult. Unlike most adjectives, the word 'undefined' does not provide any information about a particle's state; instead, it actually negates the possibility of having any information about that state. Nevertheless, this is the logical conclusion of the experiments that were used to test Bell's inequality.

It is ironic that the EPR paper, which was intended to show that quantum mechanics made an absurd prediction, became the spur for Bell to devise a physical test of that prediction. Thus, despite Einstein's protestations that quantum mechanics implies the existence of a seemingly impossible spooky action at a distance, such spookiness does exist.

1.3. The Double-Slit Experiment

The *double-slit experiment* is famous because it provides an unequivocal demonstration that light behaves like a wave. But the importance of the double-slit experiment extends far beyond that demonstration because, as Richard Feynman said in 1966:

> *In reality, it contains the **only** mystery ... In telling you how it works, we will have told you about the basic peculiarities of all quantum mechanics.*

Based on work first described by Young in 1802, the experimental apparatus consists of two vertical slits and a screen, as shown in Figure 1.2. Light emanating from each slit interferes with light from the other slit to produce an *interference pattern* on the screen. This seems to prove conclusively that light consists of waves, even though the only entities detected at the screen are individual particles, which we now call photons. The bright regions in the interference pattern correspond to areas where photons land with high probability and the dark regions to areas where photons land with low probability.

Remarkably, the same type of interference pattern is obtained even when the light is made so dim that only one photon at a time reaches the screen, as shown in Figure 1.3. With such a low photon rate, it can take several weeks for the interference pattern to emerge. However, the very fact that an interference pattern emerges at all implies that even a single photon behaves like a wave. This, in turn, seems to imply that each photon passes through both slits at the same time, which is clearly nonsense. Only a wave can pass through both slits, but only a photon can hit a single point on the screen. It is as if the wave acts as a guide, or *pilot wave*, which defines the probability that the photon will land at each point on the screen. However, this single-photon interference behaviour is not the most remarkable feature of the double-slit experiment.

If a light detector is used to measure which slit each photon passes through then the interference pattern is replaced by a diffraction envelope, as if the light consists of two streams of particles which spread out and merge over time. Crucially, a diffraction envelope is exactly what is observed if the waves from different slits are prevented

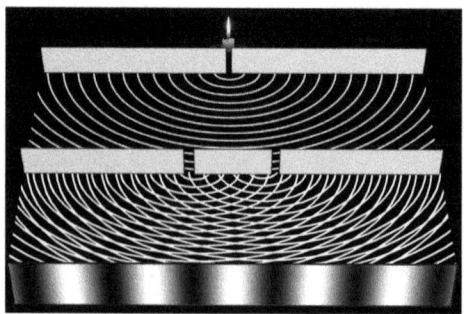

Figure 1.2: The double-slit experiment. Waves travel from the source (top) until they reach the first barrier, which contains a slit. A semi-circular wave emanates from the slit until it reaches the second barrier, which contains two slits. The two semi-circular waves emanating from these slits interfere with each other, producing peaks and troughs along radial lines that form an interference pattern on a screen (bottom).

from interfering with each other. This can be achieved by using a photographic plate to record photons as first one slit is opened on its own and then the other slit is opened on its own. In this case, the image captured by the photographic plate is the simple sum of photons from each slit (i.e. a diffraction envelope), with the guarantee that photons from the two slits could not possibly have interfered with each other. Thus, we can choose to use a light detector to find out which slit each photon passed through, but as soon as we do this, we see a diffraction envelope instead of the interference pattern. Despite many ingenious experiments, any attempt to ascertain which slit each photon passed through forces the light to stop behaving like a wave (which yields an interference pattern) and to start behaving like a stream of billiard balls (which yields a diffraction envelope). The transformation from wave-like behaviour to particle-like behaviour is intimately related to *Heisenberg's uncertainty principle*, as explained in the next section. However, even this *wave–particle duality* is not the most remarkable feature of the double-slit experiment.

If light is replaced by a beam of electrons then an interference pattern is obtained, as if the electrons were waves (see Figure 4.2, which looks like Figure 1.3). This is especially surprising, because only a few years before this result was obtained in 1927, electrons were thought to behave exactly like miniature billiard balls. Just as light can be forced to behave like billiard balls, so electrons can be forced to behave like waves.

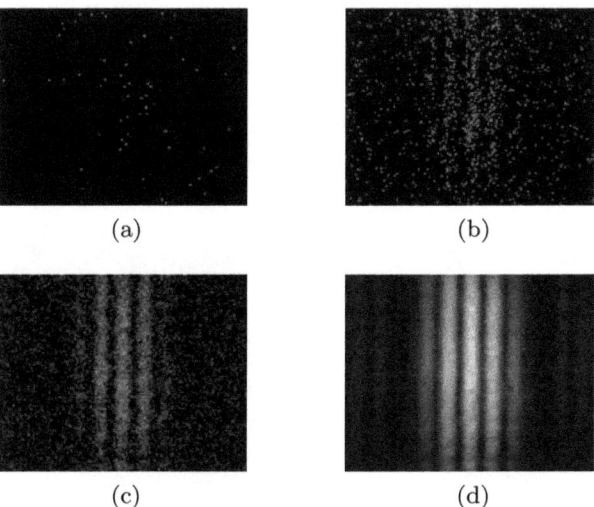

(a)   (b)

(c)   (d)

Figure 1.3: Emergence (a→d) of an interference pattern in a double-slit experiment[8], where each dot represents a photon.

The radical idea that electrons could behave like waves was proposed by de Broglie (pronounced *de Broil*) in 1923. In fact, de Broglie made the far more radical proposal that all matter has wave-like properties, now referred to as *matter waves*. Since that time, the double-slit experiment has been used to demonstrate the existence of matter waves using whole atoms and even whole molecules of buckminsterfullerene[2] (each buckminsterfullerene molecule is about 1 nanometre in diameter and contains 60 carbon atoms).

1.4. Heisenberg's Uncertainty Principle

Heisenberg's uncertainty principle states that for certain pairs of quantities, such as the position and *momentum* of a particle, it is not possible to know the values of both quantities with absolute certainty. Specifically, Heisenberg proved that the more certain we are about the position of a particle, the more uncertain we are about its momentum, and vice versa. In terms of classical physics, momentum is just mass × velocity, which allows Heisenberg's uncertainty principle to be stated in more familiar terms: for a particle with a given mass, the more certain we are about its position, the more uncertain we are about its velocity, and vice versa.

Heisenberg's uncertainty principle was originally formulated in 1927 using the German word *Ungenauigkeit*, which means *inexactness*, *indeterminacy* or *vagueness*, rather than uncertainty. In other words, Heisenberg proved that the reason we cannot know both position and momentum exactly is not because we cannot measure each of them precisely, but because these physical quantities has an irreducible degree of uncertainty, so that *in principle* they cannot both be known exactly.

At this early stage of the book, some degree of understanding can be achieved only by recourse to an analogy. For example, Schrödinger summarised the distinction between inexact measurements of well-

Figure 1.4: Heisenberg's uncertainty principle represents a perfectly focussed picture of a cloud, not an out-of-focus picture of a cat.

defined values and exact measurements of undefined values by saying[31]: "There is a difference between a shaky or out-of-focus photograph and a snapshot of clouds and fog banks." In other words, there is a world of difference between a perfectly focussed picture of a poorly defined object (like a cloud) and an out-of-focus picture of a well-defined object (like a cat). The uncertainty principle represents a perfectly focussed picture of a cloud, not an out-of-focus picture of a cat (Figure 1.4). As we shall see, the pivotal role of Heisenberg's uncertainty principle in quantum mechanics cannot be overstated.

1.5. Schrödinger's Cat

From the earliest days of quantum mechanics, physicists have made use of 'thought experiments' (*Gedankenexperiment*). A famous example of a thought experiment is Schrödinger's cat (Figure 1.5). As Schrödinger wrote[31] in 1935, "one can even set up quite ridiculous cases", such as applying superposition to ordinary objects, like cats.

The equipment required to execute Schrödinger's thought experiment (and possibly his cat) is a box, which contains a cat, a piece of radioactive uranium, a Geiger counter, a hammer, and a glass vial of poison gas. The Geiger counter is next to the piece of uranium, which is so small that exactly one of its atoms decays every two hours, on average. If an atom decays, it emits a particle that is detected by the Geiger counter, which activates a solenoid, allowing the hammer to fall onto the glass vial. When the hammer breaks the vial, the poison gas escapes, killing the cat. Because the uranium emits exactly one particle every two hours (on average), there is a 50% chance that the cat will be dead after one hour. Common sense dictates that the state of the cat is independent of whether or not anyone measures the state of the cat in the box, but quantum mechanics has a habit of disregarding common sense.

Figure 1.5: Schrödinger's cat — wanted, dead or alive (or both).

The key point is that uranium atoms decay at random times. Specifically, after one hour, we can consider every uranium atom to be in a superposition of states, so that the piece of uranium is in a superposition of two *equally probable states*, one in which it did emit a particle and one in which it did not. And if we (naively) accept that every object in the box consists of particles and that each of these also behaves like a wave, it follows that everything inside the box (including the cat) exists in a superposition of states (even though there are reasons why we should not accept this; see Section 7.3). For the uranium, this is a superposition in which it merely did, and did not, emit a particle. But for the cat, it is a superposition in which it did, and did not, shuffle off this mortal coil.

When the box is opened and the state of the cat is measured, this superposition is forced to adopt one of its two possible physical states. In one state, no particle was emitted, and the cat is alive; in the other state, a particle was emitted, and the cat is dead. The act of measuring the cat's state forces the wave to adopt a definite value, and this forcing is called the *collapse* of the superposition of waves. According to some interpretations, the collapse of the superposition of waves can be induced by something as trivial as simply looking into the box, because the act of looking can be construed as a measurement of the cat's state.

On the one hand, everyday experience suggests that the result of this experiment is just not credible, or even possible. On the other hand, quantum mechanics requires that such a result is not only possible but inevitable. Crucially, the results of experiments (in which no cats were harmed) are consistent with quantum mechanics.

In designing his thought experiment, Schrödinger hoped to show that it is nonsensical to apply quantum mechanics to everyday, macroscopic objects, by demonstrating that the idea of superposition can plausibly be applied to particles but not to cats. However, rather than closing down any further debate, Schrödinger's thought experiment was interpreted as evidence that quantum mechanics does have consequences for macroscopic objects, especially cats.

1.6. Interpreting Quantum Mechanics

Isaac Newton (1642–1726) famously derived equations that work extraordinarily well in describing how planets orbit the sun and how apples fall from trees, so these equations have a clear physical interpretation in terms of everyday experience. The equations of quantum mechanics also work well, but unlike Newton's equations they do not have any obvious physical interpretation.

For philosophers, the monumental achievements of quantum mechanics are both a blessing and a curse. They are a blessing because they provide the raw material for endless debates regarding the nature of reality, and they are a curse because the language of mathematical physics cannot always be translated into any spoken language. As one of the pioneers of quantum mechanics said, "We must be clear that when it comes to atoms, language can be used only as in poetry" (Bohr, around 1920, as later recalled by Heisenberg).

For physicists, the equations of quantum mechanics provide an account of the physical world that is without equal in terms of pure numerical accuracy. But physicists do not judge a theory on accuracy alone. Like all scientists, physicists seek a coherent, detailed theory that is both numerically accurate and intuitively satisfying. However, as we shall see in Chapter 7, quantum mechanics is the most successful and also the most counter-intuitive theory imaginable. Precisely because the equations of quantum mechanics work so well, a common response (originally stated by David Mermin) is to "shut up, and calculate".

1.7. Quantum Horizons

Quantum mechanics is currently being harnessed in many more ways than can be covered in this tutorial book. Without meaning any disrespect to other research areas within quantum mechanics, three of the most interesting developments are described briefly here.

Quantum Computers. The currency of conventional computers is the *binary digit*, or *bit*, which can adopt the state 0 or 1. By analogy, the currency of *quantum computers* is the *quantum bit*, or *qubit*. Unlike a binary digit, a qubit is a quantum state that behaves as if it is in a superposition of 0 and 1 states simultaneously. This superposition has the potential to endow quantum computers with enormous computational power, well beyond the limits of any conventional computer that could ever exist. However, the technical obstacles to building a quantum computer seem about as formidable as the problem of designing and implementing algorithms that could run on a quantum computer. Indeed, there is some controversy over whether the machines currently available actually perform quantum computations and, even if they do, whether they provide any practical advantage over conventional computers.

Quantum Cryptography. This relies on the fact that any attempt to measure a quantum system which is in a superposition of states instantly forces that system to adopt a particular state. With conventional cryptography, an encrypted message can be intercepted without the receiver being aware of it. Because quantum cryptography relies on the messaging system remaining in a superposition of states, if any interloper attempts to interfere with the system then this will be detected by the receiver. The first effective use of quantum cryptography was published[24] in 2019.

Quantum Biology. Given that quantum mechanics underpins all physical processes, we would expect that the biochemistry of living organisms relies on quantum mechanical effects. The idea of *quantum biology*[1;22] has been discussed since the birth of quantum mechanics [23;32], and it now has a research centre at the University of Surrey.

It is early days for quantum biology, and reactions from the scientific community have ranged from enthusiasm to indifferent shrugs. The enthusiasts welcome quantum biology as an approach to exploring how evolution has utilised quantum mechanical effects for photosynthesis, long-range navigation and enzyme function (for example). The shruggers regard quantum biology as nothing new, saying that it is obvious that biology relies on quantum mechanics. Even though the shruggers are self-evidently (and even tautologically) correct, knowing exactly *how* biology relies on quantum mechanics promises insights that could never be acquired by knowing only *that* biology relies on quantum mechanics.

1.8. What Makes Quantum Mechanics Hard?

Quantum mechanics is hard for two reasons. The first is that the physical phenomena which gave rise to quantum mechanics represent a fundamental contradiction of our everyday experience of the world.

The second reason it that there are three equivalent theories of quantum mechanics, developed almost simultaneously in 1925–26 by Heisenberg, Schrödinger and Dirac (Figure 1.6), with each theory having its own mathematical notation (see Chapter 8). Heisenberg's *matrix mechanics* was the first of the three theories to be proposed, but because most physicists at the time (including Heisenberg) were unfamiliar with matrices, it has never been widely adopted. Schrödinger's *wave mechanics* was formulated in terms of the differential equations that govern the behaviour of waves. These types of equations were the

(a) Heisenberg

(b) Schrödinger

(c) Dirac

Figure 1.6: Authors of three equivalent theories of quantum mechanics.

standard tools of physics at the start of the 20th century, especially since Maxwell had recently defined his enormously successful theory of electromagnetism in terms of waves. Finally, Dirac proved that the formulations of quantum mechanics by Heisenberg and Schrödinger are equivalent to Dirac's *transformation mechanics*, which introduced the *bra-ket* vector notation.

It is therefore not surprising that novices become confused when they are confronted by the counter-intuitive experimental findings of quantum mechanics, explained using a panoply of different mathematical notations. Whilst it is not possible to eliminate the fundamental strangeness of quantum mechanics, it is possible to avoid less intuitive mathematical formulations. Accordingly, the account given in this book ignores Heisenberg's matrix mechanics and Dirac's bra-ket notation, and instead relies on the mathematics of waves, which naturally leads to Schrödinger's formulation of quantum mechanics. To ensure that this account connects with our experience of the physical world, the development of quantum mechanical wave equations is done in the context of sound waves generated by a familiar object, the guitar.

1.9. Summary

The stark strangeness of quantum mechanics is epitomised by the comparison of entangled particles (non-locality, or Einstein's spooky action at a distance) with the double-slit experiment. Whereas entanglement entails that two particles can behave as if they are one particle, the double-slit experiment (using one particle at a time) demonstrates that one particle can behave as if it is many particles that somehow interfere with each other to produce an interference pattern.

In this Alice-in-Wonderland world, it is as if two particles are one and one particle is many — except that the physical world is not an Alice-in-wonderland world; its laws are not arbitrary or magical. However, as has been discovered repeatedly throughout the history of science, the laws of physics can seem illogical if we do not have the correct conceptual tools to understand them. As we shall see, a vital tool for understanding the idiosyncrasies of the double-slit experiment is Heisenberg's uncertainty principle. But no such tool has yet been discovered for understanding entanglement, despite the fact that quantum mechanics enables certain aspects of non-local behaviour to be predicted with exquisite accuracy.

The pioneering physicists Planck, Einstein, Bohr, Heisenberg, Dirac and Schrödinger developed quantum mechanics through a combination of deep insights and imaginative leaps, to reveal a menagerie of wonders and paradoxes. Using the maps drawn with such precision by these pioneers, along with the statistical compass that they perfected, we can retrace their footsteps through the quantum menagerie.

Chapter 2

Planck's Act of Desperation

> ...*what I did can be described as simply an act of desperation.*
> Planck M, 1931.

2.1. Introduction

In 1900, Max Planck was 42 years old, an established physicist in a world where physics could rightly claim to be the most successful of all branches of science. Indeed, physics was so successful that, in his youth, Planck had been advised by his professor (Philip von Jolly) to study something else because all the fundamental laws had already been discovered[21]. The prevailing view was that physics was almost complete, as succinctly stated by one of Planck's famous contemporaries:

> *most of the grand underlying principles have been firmly established... the future truths of Physical Science are to be looked for in the sixth place of decimals.*
> Michelson A, 1894.

Ironically, just three years later, Michelson was responsible (with Morley) for establishing that the speed of light, as measured by any observer, is independent of the observer's speed. This historic result played a key role in supporting Einstein's 1905 *special theory of relativity*.

Having ignored Jolly's advice for over 20 years, on the 14th of December in 1900, Planck presented a paper[28] to the German Physical Society that laid the foundations for the *quantum revolution* in physics. Planck was awarded a Nobel Prize for this work in 1918.

2.2. Classical Physics

The enormous success of classical physics in explaining various phenomena, from the orbits of planets to the propagation of electromagnetic waves, depends on an assumption so self-evident that it seems almost banal: *the amount of energy associated with any given object or wave can adopt any value in a continuous range.*

In essence, Planck violated the assumption of continuity by showing that the interaction between light radiation and matter involves discrete packets, or *quanta*. In principle, this can apply to something as small as a photon of light or as large as a pendulum, or even a planet.

For example, a pendulum swings back and forth at a constant rate, even though the extent of its swing (its amplitude) decreases over time. It seems obvious that the pendulum can adopt any swing amplitude, but what is obvious is not necessarily true. Under certain circumstances, a pendulum with a given mass can have only certain swing amplitudes, which differ from each other by a fixed, step-wise increment. The size of each step depends on *Planck's constant*, a number so small that the differences between swing amplitudes cannot easily be measured. However, just because they are hard to measure does not mean they do not exist. Before we can understand this, we need to know about blackbody radiation.

2.3. Blackbody Radiation

We now know that the universe consists of matter and waves, with energy acting as the currency for converting from one to the other. In Planck's day, it was known that light is a form of electromagnetic radiation and that it consists of waves. An important property of a wave is its *wavelength*, which is the distance between consecutive peaks and is usually represented by the Greek letter λ (lambda), as shown in Figure 2.1. If the wave travels at a speed of c m/s then its *frequency* is the number of wavelengths that pass a fixed point in one second, represented by the Greek letter ν (nu). Given that $c = \nu\lambda$, the frequency is $\nu = c/\lambda$ with units of 1/second or *hertz* (Hz).

Every object emits electromagnetic radiation across a range of wavelengths. In the case of objects hot enough to glow red, the dominant wavelength is extremely small. Visible light has a wavelength between 450×10^{-9} metres (violet/blue) and 700×10^{-9} metres (red), where

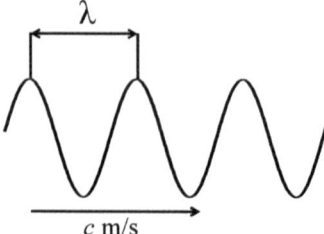

Figure 2.1: A sinusoidal wave with wavelength λ travelling at speed c m/s has a frequency of $\nu = c/\lambda$ Hz.

10^{-9}m is one nanometre (nm). For example, since $c = \nu\lambda$ and the speed of light is $c = 3 \times 10^8$ m/s, the frequency of blue light is

$$\nu = c/\lambda \qquad (2.1)$$

$$= \frac{3 \times 10^8 \text{ m/s}}{450 \times 10^{-9} \text{ m}} \qquad (2.2)$$

$$= 666 \times 10^{12} \text{ Hz,} \qquad (2.3)$$

where 10^{12} hertz is one terahertz (THz).

To study radiation under controlled conditions, physicists use an oven at a constant temperature, with a small hole in one side so that the radiation emitted from within the oven can be measured. This radiation has a range of wavelengths, which defines a characteristic *spectrum* of *blackbody radiation*, shown in Figure 2.2. By 1900, it was known that an oven at a given temperature emits blackbody radiation and that the spectrum has the same characteristic form irrespective of the shape of the oven used to produce the radiation. These facts suggested there was something fundamental about the blackbody spectrum. However, before 1900, no equation had been discovered that would fit the blackbody spectra shown in Figure 2.2, because the physics underpinning such spectra was not understood.

In historical terms, the problem of blackbody radiation was not considered hugely important; it was just one of a handful of unremarkable problems waiting to be solved. Given the successful track record of 19th-century classical physics, it seemed reasonable to assume that

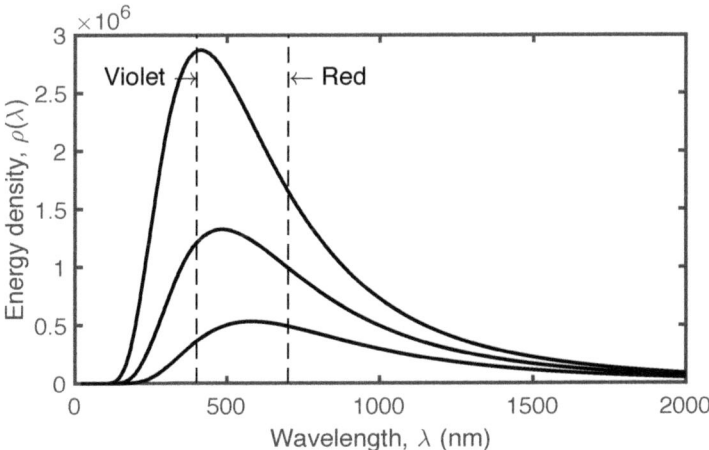

Figure 2.2: Blackbody spectra at 7000 K (top curve), 6000 K (middle curve) and 5000 K (bottom curve). The dashed vertical lines correspond to the wavelengths of violet and red light.

the problem of blackbody radiation would also be solved before long. And so it was. However, in the process of solving the problem, Planck unwittingly made a crack in classical physics, through which quantum physics would seep like molten lava.

2.4. Planck's Inspired Guesswork

The solution found by Planck represented (in his own words) the hardest work he had ever done. During a brief intense period, Planck's struggles wrenched him from the familiar world of classical physics to the threshold of the brave new world of quantum mechanics. The intensity of those struggles is implicit in the inscrutable and, at times, opaque account of his results, published in 1900. As Planck said later, this paper was

> a result of inspired guesswork, scientific tact, sober compromise, in short, tinkering.

Indeed, it seems likely that Planck's paper was convoluted precisely because his new equation violated a fundamental assumption implicit in all of physics up to that time, namely that the energy of a particle can adopt any value in a continuous range. To skip ahead a little, energy *can* adopt any value in a continuous range, but Planck's epiphany was that the energy of blackbody radiation can only *change* by a discrete amount, or a *quantum*.

To save ourselves from the deductive agonies and cul-de-sacs experienced by Planck, we restrict the account below to a derivation that is simpler than Planck's but which still captures the flavour of Planck's ideas.

Planck's goals were (and now our goals are) to find an equation that fits the blackbody spectra in Figure 2.2, together with a physical explanation for the form of that equation.

Planck's First Step. Planck's equation consisted of just two main terms, so it was a kind of mathematical sketch, with details that would be filled in later. The blackbody spectrum shown in Figure 2.2 is defined formally as the *spectral energy density*, and Planck's initial sketch was

$$\rho(\nu) \;\;=\;\; N(\nu) \times \overline{E}(\nu) \text{ J/m}^3. \tag{2.4}$$

We will explore each term in more detail below, but, roughly speaking, $N(\nu)$ is the number of *standing waves* per cubic metre for a given frequency ν, and $\overline{E}(\nu)$ is the average energy of standing waves at that frequency. Therefore, $\rho(\nu)$ has units of energy per cubic metre.

2.5. Counting Standing Waves

Tapping a wine glass produces a pure tone, a single sinusoidal frequency, which is its natural or *resonant frequency*. In fact, every object has a resonant frequency, which is most obvious when oscillations have minimal *damping*, as in the case of the wine glass (e.g. pressing a finger on the rim of the glass dampens the oscillations). Other examples of structures that have a reasonably obvious resonant frequency are a pendulum, a swing, a flute, a guitar string, and the electromagnetic waves inside a metal oven (being metal means that no electromagnetic waves can escape). Each resonant frequency is the result of a standing wave, whose wavelength is determined by the size of the structure that constrains the wave. Because electromagnetic waves and waves on guitar strings have much in common, we begin by exploring the more familiar example of oscillations of a guitar string. When the string is vibrating, every point along its length is free to move except for the two end points, and this constrains the manner in which the string can vibrate, as shown in Figure 2.3.

Specifically, if the length of the string is L then half a wavelength fits into L, so the fundamental frequency or first *mode* has a wavelength of $\lambda_1 = 2L$. For the second mode (first harmonic), two half-wavelengths of it fit into L, so the wavelength is $\lambda_2 = L$. For the third mode, three half-wavelengths fit into L, giving a wavelength of $\lambda_3 = 2L/3$, and so on. Thus, if the number of half-wavelengths in L is exactly n then the wavelength of the nth mode is

$$\lambda_n \;=\; 2L/n \text{ m}, \tag{2.5}$$

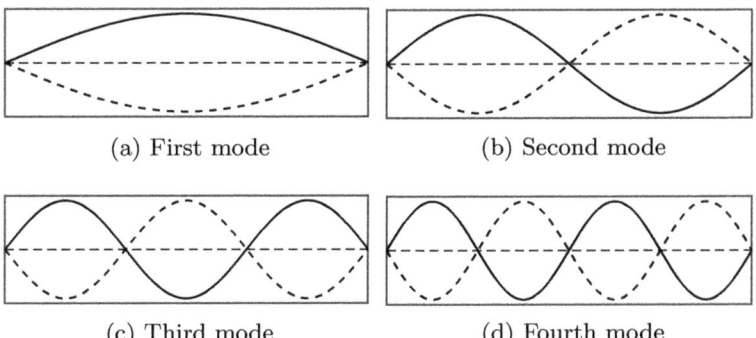

(a) First mode   (b) Second mode

(c) Third mode   (d) Fourth mode

Figure 2.3: The first four modes (sinusoidal standing waves) in a guitar string; each curve represents the amplitude at every position.

as shown in Figure 2.3. If the speed of the wave along the string is v, then the frequency is $\nu = v/\lambda$, so the frequency of the first mode is

$$\nu_1 \;=\; v/\lambda_1 \;=\; v/(2L) \text{ Hz}, \qquad (2.6)$$

and the frequency of the second mode is

$$\nu_2 \;=\; v/\lambda_2 \;=\; v/L \;=\; 2v/(2L) \text{ Hz}. \qquad (2.7)$$

Using $v/(2L) = \nu_1$ from Equation 2.6, we can express the frequency of each mode in terms of the frequency of the first mode, so

$$\nu_2 \;=\; 2\nu_1 \text{ Hz} \qquad (2.8)$$

and, in general, the frequency ν_n of the nth mode is

$$\nu_n \;=\; n\nu_1 \text{ Hz}, \quad \text{for } n = 1, 2, \ldots. \qquad (2.9)$$

The main point is that for a string fixed at both ends, the only frequencies allowed are integer multiples of the fundamental frequency, and each such frequency is associated with a standing wave. It can be shown that any other frequency effectively cancels itself out and therefore has an amplitude of zero. What have guitar strings to do with blackbody radiation? Except for a change in the speed of the wave (from v to c, the speed of light), what is true for a guitar string is also true of electromagnetic waves.

Just as a standing wave on a guitar string is static at both ends, so constraining the ends of an electromagnetic wave inside a container ensures that it is a standing wave. In both cases, it is the absence of motion at both ends that guarantees that any wave must be a standing wave. Accordingly, we can replace the length of the guitar string with the length of an oven.

Consider an oven shaped like a cube with side length L metres; then it has a volume of $V = L^3$ m^3. From Equation 2.5, the only frequencies allowed define standing waves for which an integer number n of half-wavelengths $\lambda_n/2$ fit into the length L: $n(\lambda_n/2) = L$, so

$$n \;=\; 2L/\lambda_n. \qquad (2.10)$$

Using lengthy (and tedious) counting arguments, Rayleigh showed[11] that the number of electromagnetic standing waves per cubic metre between the frequencies ν and $\nu + d\nu$ is

$$N(\nu)\, d\nu \;=\; \frac{8\pi\nu^2}{c^3}\, d\nu, \qquad (2.11)$$

and that this applies not only to cubic ovens but to ovens of any shape. It is worth emphasizing that this equation was derived using standard methods from classical physics. Rayleigh's equation for $N(\nu)$ meant that Planck had a half-sketched equation for blackbody radiation, as embodied in Equation 2.4. Substituting Equation 2.11 into Equation 2.4 gives

$$\rho(\nu) \quad = \quad \frac{8\pi\nu^2}{c^3} \times \overline{E}(\nu). \tag{2.12}$$

However, what Planck did not have at this point was a convincing expression for $\overline{E}(\nu)$. To understand how Planck arrived at his equation for $\overline{E}(\nu)$, we need to survey two other attempts at finding such an equation. Even though both of these attempts failed, each supplied vital clues that eventually enabled Planck to succeed.

2.6. The Ultraviolet Catastrophe

In 1900, Rayleigh presented an analysis based on classical physics and statistical mechanics that was self-evidently mostly wrong. Crucially, the part of his analysis that was not wrong included Equation 2.11, which specifies how to estimate $N(\nu)$ (i.e. the first term in Equation 2.4). However, following the tradition established by Boltzmann, Rayleigh assumed that all entities (e.g. frequencies) are equally common.

In essence, Rayleigh applied the *law of equipartition of energy* to standing waves. Briefly, this law states that each element of system in a state of thermal equilibrium should have the same amount of energy. For example, for a container of gas molecules in thermal equilibrium at temperature T, the average kinetic *energy per degree of freedom* is $k_B T/2$, where k_B is *Boltzmann's constant*. Boltzmann's constant is equal to 1.381×10^{-23} J/K, where K is the temperature unit of kelvins.

The number of degrees of freedom of any entity is the total number of parameters required to describe its state. For example, the position of a particle requires three parameters, which are its locations along the x-, y- and z-axes. Therefore, an oven containing two particles requires three position parameters per particle, making a total of six degrees of freedom. In general, an oven containing N particles requires three position parameters per particle, making a total of $3N$ degrees of freedom. Of course, if each particle has other properties (e.g. polarising angle) then this increases the number of degrees of freedom per particle.

If we follow Rayleigh's logic and replace molecules with sinusoidal standing waves, then we find that each standing wave has only one degree of freedom, its amplitude. Following the traditions of classical physics, Rayleigh reasoned that all standing waves are allowed to interact. This led inexorably to the incorrect conclusion that their amplitudes would reach a state of equilibrium such that standing waves at all frequencies have the same average energy (equal to $k_B T$). Thus, Rayleigh assumed that the law of equipartition would imply that the average energy of standing waves is independent of the frequency ν; specifically,

$$\overline{E}(\nu) \quad = \quad k_B T \text{ J} \qquad\qquad (2.13)$$

(skip ahead to Section 2.7 to see why this is true). Substituting Equation 2.13 into Equation 2.12 yields the spectral energy density formula

$$\rho(v) \quad = \quad \frac{8\pi\nu^2}{c^3} k_B T. \qquad\qquad (2.14)$$

Rayleigh developed this equation over a number of years with the help of Jeans, so it is known as the *Rayleigh–Jeans formula for blackbody radiation*. It provides a good fit to the data at low frequencies, as shown in Figure 2.4. However, it also predicts that energy density increases *without limit* in proportion to the square of frequency. The divergence between the Rayleigh–Jeans formula and the empirical blackbody spectrum becomes substantial towards the ultraviolet end of the visible spectrum. This rapidly increasing divergence later became known as the *ultraviolet catastrophe*.

Wien's Infrared Catastrophe. An earlier proposal by Wien in 1893 was also flawed, but in a way that was complementary to the Rayleigh–Jeans formula. Specifically, Wien's formula defined a curve that provided a reasonable fit to the blackbody spectrum at high but not at low frequencies, as shown in Figure 2.4. So, whereas the Rayleigh–Jeans formula gave rise to the ultraviolet catastrophe, by symmetry we could say that Wien's formula gave rise to an *infrared catastrophe* (though it

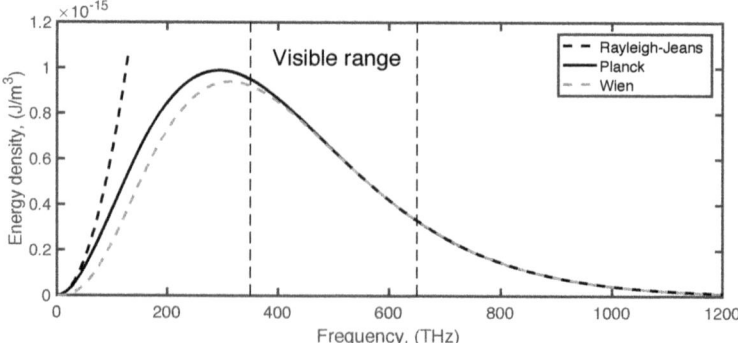

Figure 2.4: The blackbody spectrum as modelled by Planck (solid curve, Equation 2.41), by Wien (grey dashed curve, Equation 2.15) and by Rayleigh and Jeans (black dashed curve, Equation 2.14) at $T = 7000$ K.

was never called that at the time, perhaps because it isn't really very catastrophic). For completeness, *Wien's radiation law* is

$$\rho(v) = \frac{2h\nu^3}{c^2} e^{-h\nu/(k_B T)}. \tag{2.15}$$

In terms of scientific slips, Equation 2.15 was by no means Wien's worst. As one of the examiners at Werner Heisenberg's PhD viva (oral exam), Wien insisted that Heisenberg should fail because of a minor error regarding the optics of microscopes. Had the other examiners allowed this judgment to stand, it is almost certain that *Heisenberg's uncertainty principle* would be named after someone else. In the end, the examiners agreed to award this future Nobel Prize winner a PhD, but at the lowest level of pass.

Wien's radiation law was also developed by Planck, to the extent that (before 1900) it was known as the *Wien–Planck law*. Remarkably, while working on the Wien–Planck law, Planck had already deduced that a new universal constant must exist, whose value is equal to (what would later become) Planck's constant (see Nelson in Further Reading).

Avoiding Both Catastrophes. Unlike the Rayleigh–Jeans formula, Wien's formula relied on the *Boltzmann distribution* (the exponential is the relevant term in Wien's formula). Indeed, Wien's formula is noteworthy because it includes a term that can be interpreted as a Boltzmann distribution. However, the dependence on the Boltzmann distribution was expressed very differently by Wien and by Planck — so differently that only Planck's equation does not suffer from Wien's infrared catastrophe. Therefore, to understand Planck's equation, we first need to understand the Boltzmann distribution.

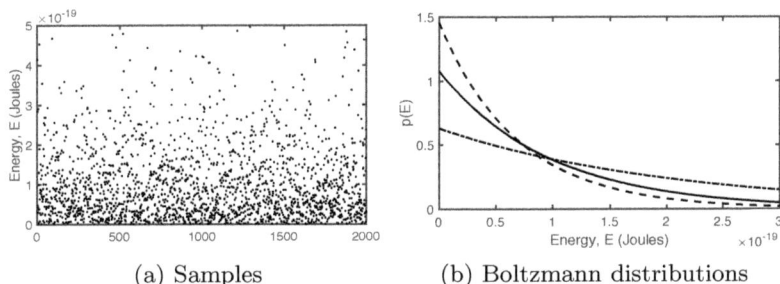

(a) Samples   (b) Boltzmann distributions

Figure 2.5: The Boltzmann distribution. (a) Each dot represents one of 2000 random samples from a Boltzmann distribution. (b) The Boltzmann distribution, Equation 2.16, at 5000 K (dashed), 7000 K (solid) and 15000 K (dot-dashed).

2.7. The Boltzmann Distribution

Consider a closed system, like a jar of gas. Empirically, different particles are found to have different speeds, which have a characteristic distribution at a given temperature. Specifically, each particle (molecule) with mass M and speed v has a kinetic energy of $E = Mv^2/2$. The probability (density) that a randomly chosen particle has energy E is given by the Boltzmann distribution

$$p(E) \;=\; \frac{1}{Z}\,e^{-E/(k_{\mathrm{B}}T)}, \tag{2.16}$$

where T is temperature in kelvins, k_{B} is Boltzmann's constant (1.38×10^{-23} J/K) and Z is called the *partition function*.

Equation 2.16 actually defines a family of distributions, with each value of T giving a different member of that family, as shown in Figure 2.5b. Basically, $p(E)$ decreases exponentially with increasing values of E, but the temperature determines how quickly $p(E)$ decreases. At low temperatures $p(E)$ falls rapidly with E, whereas at high temperatures $p(E)$ falls slowly. The partition function is so called because it is a function of temperature,

$$Z \;=\; \int_{E=0}^{\infty} e^{-E/(k_{\mathrm{B}}T)}\, dE. \tag{2.17}$$

Its purpose is to ensure that the total probability of E sums to 1, so that (for example) the mean energy is

$$\overline{E} \;=\; \int_{E=0}^{\infty} p(E)E\, dE. \tag{2.18}$$

Substituting Equations 2.16 and 2.17 into Equation 2.18, we obtain

$$\overline{E} \;=\; \frac{1}{Z}\int_{E=0}^{\infty} e^{-E/(k_{\mathrm{B}}T)}E\, dE, \tag{2.19}$$

which is the area under the curve of $p(E)E$ in Figure 2.6. For a derivation of the Boltzmann distribution, see Appendix C.

2.8. An Act of Desperation

If the energy of each standing wave at frequency ν is $E(\nu)$ and if the proportion of all standing waves with energy $E(\nu)$ is $p(E(\nu))$, then the average energy at frequency ν is

$$\overline{E}(\nu)^{\mathrm{cont}} \;=\; \int_{E(\nu)=0}^{\infty} p(E(\nu))\, E(\nu)\, dE(\nu) \;=\; k_{\mathrm{B}}T, \tag{2.20}$$

where the superscript ^{cont} indicates that energy varies continuously. Notice that, when substituted into Equation 2.12, the value $\overline{E}(\nu) = k_{\mathrm{B}}T$ yields the ultraviolet catastrophe (Equation 2.14).

Now consider an oven with fixed total energy that contains a large number of standing waves, where each standing wave vibrates at a frequency of ν Hz. If we assign continuous energy values E to these standing waves at random, we will end up with a Boltzmann distribution of wave energies as defined in Equation 2.16. Rather than standing waves, Planck imagined the walls of the oven to contain numerous tiny *oscillators*, which we would now call *photons*. Because the mathematics of standing waves and oscillators is the same, we shall proceed to study standing waves.

Planck realised he could obtain the correct (i.e. empirical) form for energy density if he treated $E(\nu)$ as a discrete variable, as opposed to the continuous variable assumed in classical physics (e.g. Equation 2.20). Specifically, he needed to make energy increase in steps that are proportional to the standing wave frequency:

$$E(\nu) \quad = \quad h\nu,\ 2h\nu,\ \ldots, \tag{2.21}$$

where the constant h is known as *Planck's constant*. This defines a frequency-dependent *quantum* of energy

$$\Delta E(\nu) \quad = \quad h\nu. \tag{2.22}$$

Given today's knowledge of photons, we can interpret this to mean that there can only be an integer number n of photons (oscillators) at frequency ν, each of which has energy $h\nu$, so that the total energy at frequency ν is

$$E(\nu) \quad = \quad nh\nu. \tag{2.23}$$

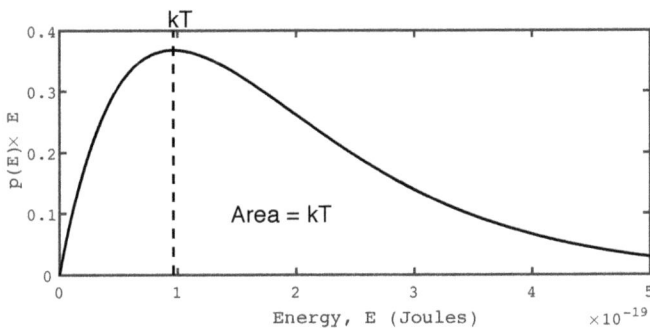

Figure 2.6: The quantity $p(E) \times E$ as a function of E, where $p(E)$ is the Boltzmann distribution shown in Figure 2.5b (Equation 2.16).

In this case, the integrand in Equation 2.20 is non-zero only at discrete values of $E(\nu)$. We can evaluate Equation 2.20 at discrete values of $E(\nu)$ by employing the *Dirac delta function* as follows:

$$\int_E \delta(E(\nu) - nh\nu)\, dE = \begin{cases} 1 & \text{if } E(\nu) - nh\nu = 0, \\ 0 & \text{if } E(\nu) - nh\nu \neq 0. \end{cases} \qquad (2.24)$$

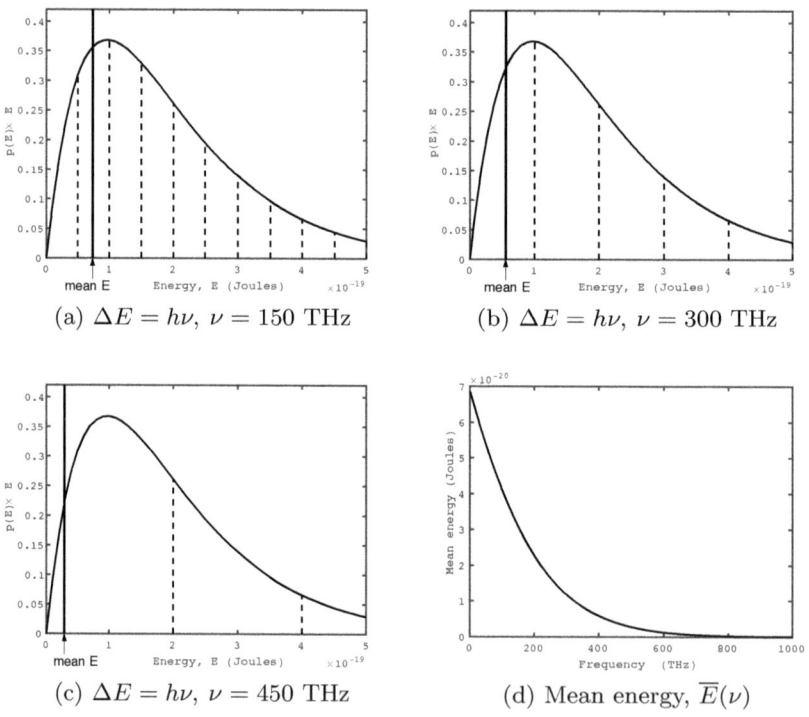

Figure 2.7: The energy of oscillators at different discrete frequencies. The mean energy $\overline{E}(\nu)$ at frequency ν is a weighted average of energy values $E_n(\nu) = nh\nu$ with weights $p(E_n(\nu))$, for $n = 1, \ldots, \infty$ (dashed lines). The mean E value $\overline{E}(\nu)$ at frequency ν is indicated by the thick lines in (a)–(c).

(a) At a low frequency of 150 THz, successive $E_n(\nu)$ values are separated by small intervals $\Delta E = h\nu$, so the mean $\overline{E}(\nu)$ (Equation 2.27) is dominated by high values on the curve $p(E_n(\nu))E_n(\nu)$.

(b) At $\nu = 300$ THz, $\overline{E}(\nu)$ is dominated by lower values on the curve.

(c) At $\nu = 450$ THz, $\overline{E}(\nu)$ is dominated by low values on the curve.

(d) Consequently, $\overline{E}(\nu)$ decreases with frequency (Equation 2.40).

In words, this integral evaluates to 1 only if $E(\nu)$ is an exact integer multiple of $h\nu$ (i.e. $E(\nu) = nh\nu$). Accordingly, Equation 2.20 becomes

$$\overline{E}(\nu) \;=\; \int_{E=0}^{\infty} \delta(E(\nu) - nh\nu)\, p(E(\nu))\, E(\nu)\, dE, \qquad (2.25)$$

which can be written as the sum

$$\overline{E}(\nu) \;=\; \sum_{n=0}^{\infty} p(E_n(\nu)) E_n(\nu) \qquad (2.26)$$

where $E_n(\nu) = nh\nu$, and so Equation 2.26 can be written as

$$\overline{E}(\nu) \;=\; \sum_{n=0}^{\infty} p(nh\nu)\, nh\nu \;=\; \mathrm{E}[nh\nu], \qquad (2.27)$$

where the upright $\mathrm{E}[\,\cdot\,]$ stands for mean, *expected value* or *expectation*.

At low frequencies, consecutive values of $nh\nu$ are separated by small intervals of size $\Delta E(\nu) = h\nu$, as in Figure 2.7a. This means that the function $p(E(\nu))E(\nu)$ in the integrand of Equation 2.20 is sampled at small intervals by the integrand of Equation 2.25; consequently, the sum in Equation 2.27 is a little less than the mean of $k_\mathrm{B}T$ obtained with the integral in Equation 2.20, as indicated by the thick line in Figure 2.7a. Indeed, if the frequency is sufficiently low, $\Delta E(\nu)$ approaches zero (which approximates Equation 2.20) and therefore

$$\mathrm{E}[nh\nu] \;=\; k_\mathrm{B}T \;\approx\; 1 \times 10^{-19}\ \mathrm{J}, \qquad (2.28)$$

where we have assumed $T = 7000\,\mathrm{K}$ here. At higher frequencies, $\Delta E(\nu) = nh\nu$ becomes larger, so the function $p(E(\nu))E(\nu)$ is sampled at larger intervals; hence, the sum over all such products shrinks towards zero, as shown by the thick vertical lines in Figures 2.7a–c. A summary of how the mean $\overline{E}(\nu)$ decreases with frequency is shown in Figure 2.7d.

To understand the number of Planck's oscillators (equivalently, the number of standing waves or photons) at each energy level, we assume that each oscillator requires a minimum amount of energy to become active. Accordingly, the proportion of oscillators with a given energy level falls rapidly with increasing frequency, as in Figure 2.5b; this in turn implies that the mean energy E at each frequency decreases as in Figure 2.7d. Next, we derive an equation that describes this behaviour. Substituting Equation 2.23 into Equation 2.16 gives

$$p(E_n(\nu)) \;=\; \frac{1}{Z_\nu}\, e^{-nh\nu/(k_\mathrm{B}T)}, \qquad (2.29)$$

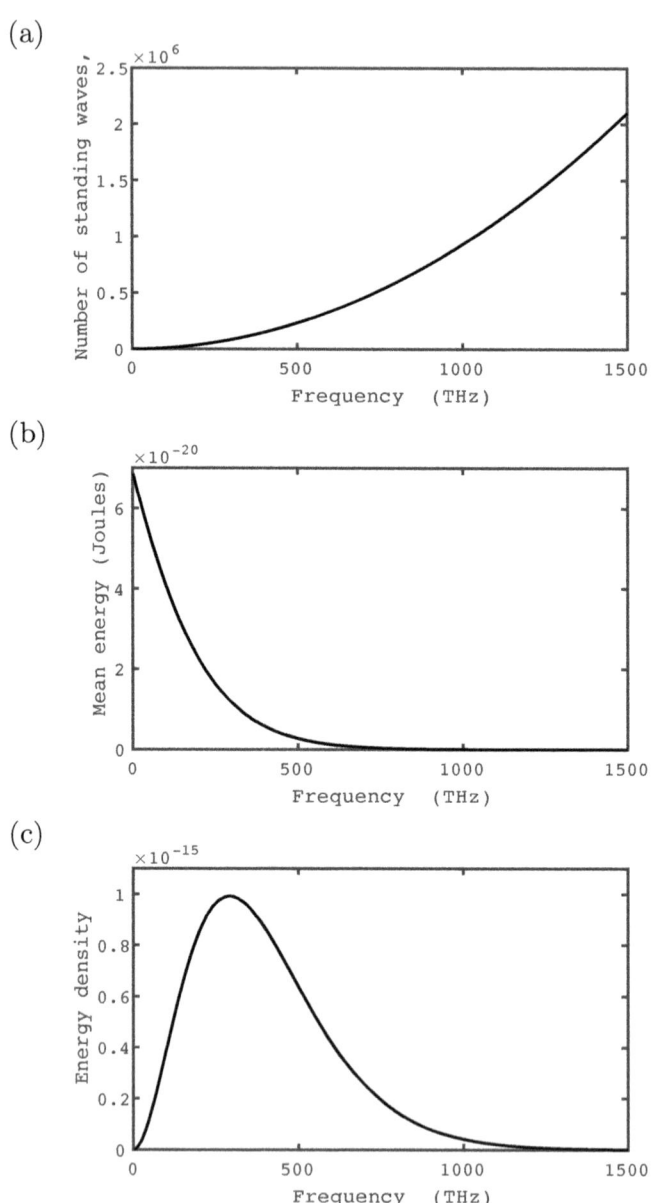

Figure 2.8: Constructing Planck's energy density spectrum.
(a) The number $N(\nu)$ of standing waves at each frequency (Eq. 2.11).
(b) Mean energy $\overline{E}(\nu)$ as a function of frequency (Eq. 2.40) at T=5,000K.
(c) Energy density $\rho(\nu) = N(\nu)\overline{E}(\nu)$ (Eqs 2.4 and 2.41).

where Z_ν is a discretized version of the partition function Z,

$$Z_\nu = \sum_{n=0}^{\infty} e^{-nh\nu/(k_{\mathrm{B}}T)}. \tag{2.30}$$

Substituting Equation 2.29 into Equation 2.27 yields

$$\overline{E}(\nu) = \frac{1}{Z_\nu} \sum_{n=0}^{\infty} nh\nu\, e^{-nh\nu/(k_{\mathrm{B}}T)} \tag{2.31}$$

$$= \frac{\sum_{n=0}^{\infty} nh\nu\, e^{-nh\nu/(k_{\mathrm{B}}T)}}{\sum_{n=0}^{\infty} e^{-nh\nu/(k_{\mathrm{B}}T)}}. \tag{2.32}$$

This can be simplified as follows (or skip to Equation 2.40). First, to simplify notation, temporarily set $x = e^{-h\nu/(k_{\mathrm{B}}T)}$ so that

$$\overline{E}(\nu) = \frac{h\nu \sum_{n=0}^{\infty} nx^n}{\sum_{n=0}^{\infty} x^n} \tag{2.33}$$

$$= h\nu \times \frac{0x^0 + 1x^1 + 2x^2 + 3x^3 + \cdots}{x^0 + x^1 + x^2 + \cdots} \tag{2.34}$$

$$= h\nu \times \frac{x + 2x^2 + 3x^3 + \cdots}{1 + x + x^2 + \cdots} \tag{2.35}$$

$$= h\nu \times \frac{x(1 + 2x + 3x^2 + \cdots)}{1 + x + x^2 + \cdots}. \tag{2.36}$$

We then make use of the series expansions

$$1/(1-x) = 1 + x + x^2 + x^3 + \cdots, \tag{2.37}$$
$$1/(1-x)^2 = 1 + 2x + 3x^2 + \cdots \tag{2.38}$$

to obtain

$$\overline{E}(\nu) = \frac{h\nu x}{1-x} = \frac{h\nu}{x^{-1} - 1}. \tag{2.39}$$

Upon reinstating $x = e^{-h\nu/(k_{\mathrm{B}}T)}$, we get

$$\overline{E}(\nu) = \frac{h\nu}{e^{h\nu/(k_{\mathrm{B}}T)} - 1}. \tag{2.40}$$

Just to be clear, Equation 2.40 gives the average energy $\overline{E}(\nu)$ at each frequency ν. This average is taken over every possible number n of photons, where the probability that n photons have frequency ν decreases exponentially with n.

Finally, substituting Equation 2.40 into Planck's Equation 2.4 yields the energy density for *Planck's blackbody spectrum*:

$$\rho(\nu) \quad = \quad \frac{8\pi\nu^2}{c^3} \frac{h\nu}{e^{h\nu/(k_{\mathrm{B}}T)} - 1}. \tag{2.41}$$

The contribution of each term on the right-hand side of Equation 2.41 can be seen in Figure 2.8. The first term increases with the square of frequency as in Figure 2.8a. For all practical purposes, the second term decreases exponentially with frequency as in Figure 2.8b. The product of these curves at corresponding values of frequency yields Planck's blackbody spectrum shown in Figure 2.8c. The constant h is now known as *Planck's constant* and has a value of 6.626×10^{-34} J s (joule seconds).

When Planck's constant is multiplied by an electromagnetic wave with frequency ν Hz (i.e. s^{-1}), the result has units of energy (joules, or J), so Planck's constant provides a link between frequency and energy. Indeed, if we express Planck's constant in terms of frequency, $h = 6.626 \times 10^{-34}$ J/Hz, we see that increasing the frequency from ν Hz to $\nu + 1$ Hz requires an additional energy of $E = h(\nu + 1) - h\nu = h$ J.

Hypothetically, if Planck's constant were allowed to shrink towards zero then $E[\nu]$ would approach $k_{\mathrm{B}}T$ (Equation 2.20), as in the Rayleigh–Jeans formula (Equation 2.14). Therefore, if Planck's constant did not have a finite non-zero value then the simple act of lighting a fire would generate an infinite amount of energy.

In a series of papers between 1911 and 1913, Einstein and Stern found that the average energy of a standing wave has a *zero-point* or *ground-state* energy (*Nullpunktsenergie*) of $h\nu/2$. This means that Equation 2.40 should include a correction term of $h\nu/2$ (see Section 4.5).

2.9. Summary

Planck was known to be a conservative character, trained in the traditions of classical physics defined in terms of a strictly Newtonian universe. His famous act of desperation was just that — he thought of it as a mathematical trick which would be replaced by a more complete classical theory. But despite his best efforts to find a non-quantum alternative, Planck eventually had to concede that Newton's universe was just an approximation to an underlying quantum universe. Thus, it seems that this most unrevolutionary scientist unwittingly lit the fuse to the most fundamental revolution in the history of physics.

Chapter 3

Einstein's Unreasonable Reality

Thou canst not stir a flower
Without troubling of a star.
Thompson F, 1913.

3.1. Introduction

The history of quantum mechanics is replete with thought experiments designed to test its validity. One of the most famous of these is the so-called *EPR paradox*, published in 1935 by Albert Einstein, Boris Podolsky and Nathan Rosen[10]. In essence, the EPR paper showed that quantum mechanics predicts that particles can communicate with each other at faster-than-light speed. This is the famous *spukhafte Fernwirkung*, or *spooky action at a distance*. More importantly, and more bizzarely, the EPR paper showed that the very existence of a definite value for a given physical quantity depends on the act of measurement. Einstein objected to this, famously declaring: "No reasonable definition of reality could be expected to permit this." However, almost a century after the EPR paper, it has become apparent that Einstein was correct in one respect: quantum mechanics really does represent an unreasonable definition of reality.

As mentioned in Chapter 1, the phrase *spooky action at a distance* was coined by Einstein and was intended as a direct attack on quantum mechanics. It refers to the seemingly impossible prediction that if two particles are *entangled* then measuring the state of one particle *instantly* affects the state of the other particle, even if the particles are separated by an astronomical distance. In modern terminology, the notion of action at a distance is called *non-local behaviour*.

A fairly obvious (but wrong) explanation for spooky action at a distance is that the two particles of an entangled pair are analogous to the left and right hands of a glove; if one particle is measured and found to be a right glove then the other particle must be a left glove. It is as if both members of a pair of particles carry a copy of the same set of instructions or programs, which are essentially *hidden variables*. These hidden variables ensure that the outcome of any measurement made on one particle of a pair logically implies the value of a measurement made on the other particle. In effect, the hidden variables make it look as if two particles can communicate with each other, but this amounts to the equivalent of a classical physics card trick.

The problem was that it seemed as if no experiment would be able to distinguish between this *hidden-variable hypothesis* and the instantaneous communication predicted by the EPR paper. In 1964, however, John Bell published an ingenious theorem[4] that provided a test for the hidden-variable hypothesis. This theorem defines *Bell's inequality*, which is violated if the hidden-variable hypothesis is false.

In effect, the EPR paper provided the impetus for the development of a vital test of quantum mechanics described in Bell's paper. Crucially, numerous experiments designed to test Bell's inequality have found that it is violated, and therefore that the hidden-variable hypothesis is false. However, the most remarkable implication of the violation of Bell's theorem is not merely that two particles communicate with each other instantaneously when one of them is measured, but that their states cannot possibly have definite values before such a measurement is made[20].

3.2. Quantum Filters

In this chapter we explore the predictions of the EPR paper in terms of polarisation angle, so we need to know about light and polarising filters.

Light is an *electromagnetic wave* that consists of two *orthogonal* (i.e. perpendicular) oscillating components, the electric component and the magnetic component (see Figure 3.1). For our purposes, we need only consider the electric component.

For a light wave travelling in a particular direction, its electric field can oscillate along any orientation that is orthogonal to the wave's direction of travel. For example, light travelling along the x-axis can oscillate along any orientation between $\pm 90°$ in the z–y plane (in which there are a total of $180°$ of possible orientations). A *polarising filter* has a particular polarising angle α within the z–y plane, such that any light

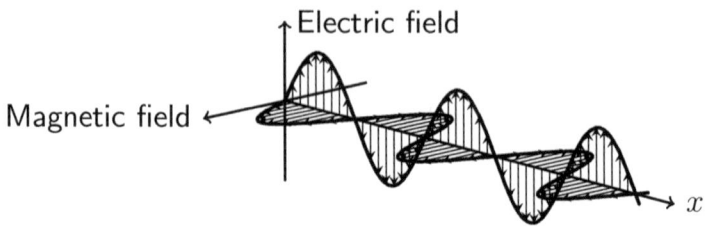

Figure 3.1: A light wave travelling in the x direction has electric and magnetic fields that oscillate in orthogonal orientations.

that passes through the filter emerges with an oscillation orientation of $\beta = \alpha$ (see Figure 3.2). This means that we can generate waves with a specific polarised angle β by passing light through a filter with a polarising angle of $\alpha = \beta$.

To avoid confusion later on, it is important to note that an average of 50% of the photons in light will pass through any given filter. One way to think about this (which is incorrect but will suffice for now) is to assume that unpolarised light consists of a mixture of polarised angles between $\pm 90°$. With this incorrect assumption, given a filter with a polarising angle of, say, $\alpha = 0°$, only photons with a polarised angle within $\pm 45°$ of $0°$ will pass through that filter. Because we assume that light has polarised angles between $\pm 90°$, about half of the waves will be within $\pm 45°$ of $0°$, so about half of them will pass through the filter.

However, as we shall see, the assumption that unpolarised light consists of a mixture of polarised angles is wrong. Indeed, most of this chapter is concerned with proving that waves in unpolarised light cannot have a definite polarised angle at all. Thus, by the end of this chapter we will arrive at the following conclusion: An entangled pair of photons can behave as if they are the same photon. Consequently, when both photons meet the same filter, either they both pass through or they are both absorbed. Moreover, each photon does not have a definite polarised angle unless it passes through a filter, so it is pure folly to try to emulate the behaviour of entangled photon pairs by assigning the same definite polarised angle to each pair (i.e. using the hidden-variable trick), because it simply cannot be done.

Three different accounts of entanglement are given in the following sections: the short version in Section 3.5, the counting of photon pairs in Section 3.6, and the hidden-variable hypothesis in Section 3.7. Sections 3.3 and 3.4 are brief but necessary diversions.

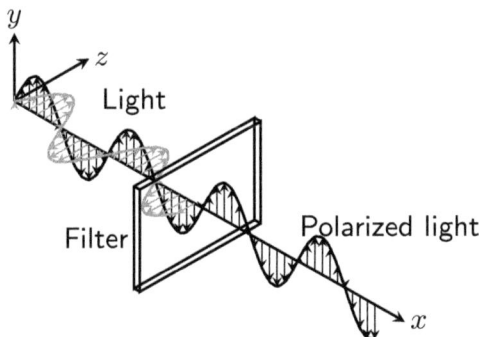

Figure 3.2: A filter with a vertical polarising angle $\alpha = 0°$. Of the two light waves with polarised angles $\beta = 90°$ (grey) and $\beta = 0°$ (black), only the wave with $\beta = \alpha = 0°$ passes through the filter.

3.3. Single Photons

Here we consider the behaviour of single photons as they attempt to pass through a pair of filters, like those depicted in Figure 3.5a. For convenience, we label the two filters B and C, with vertical polarising angles $\alpha_B = 0°$ and $\alpha_C = 0°$. If filter B is used alone then 50% of the light passes through it, and the same is true of filter C. If B is laid on top of C then the amount of light that passes through both filters depends on how much C is rotated with respect to B. If the filters are in their original orientations (i.e. $\alpha_B = \alpha_C = 0°$) then filter B allows 50% of the light through, and all of that light also passes through filter C. Specifically, if a photon passes through filter B then it also passes through filter C, and if a photon is absorbed by filter B then it would also be absorbed by filter C (e.g. if filter B is removed).

If C is now rotated so that $\alpha_C = 90°$, then 50% of the light passes through filter B (as before) but none of that light passes through filter C. As C is gradually rotated away from 90°, an increasing proportion of the light that passed through B will also pass through C. Quantum mechanics predicts that the percentage of photons that will pass through a filter C with polarising angle α_C, given that they have already passed through a filter B with polarising angle α_B, is

$$q(\mathrm{B}, \mathrm{C}) \quad = \quad (100 \times \cos^2 \gamma) \ \% \qquad\qquad (3.1)$$

where $\gamma = \alpha_B - \alpha_C$, as shown in Figure 3.3.

Before proceeding, we can greatly simplify matters by defining the *agreement rate* as the percentage of photons that pass through filter C given that they have passed through B. For example, if filters B and C are perfectly aligned (i.e. $\alpha_B = \alpha_C = 0°$) then all of the light that

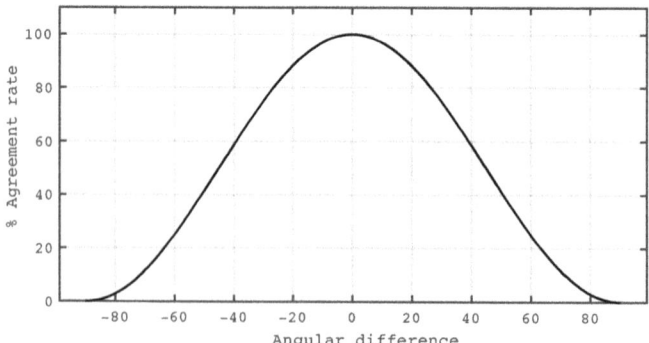

Figure 3.3: The agreement rate $q(\mathrm{B}, \mathrm{C})$, the percentage of photons that pass through filter C given that they have passed through filter B, decreases as the difference γ between the polarising angles of B and C increases (Equation 3.1).

passes through B also passes through C, so the agreement rate is 100%. At this point, we define the *quantum mechanical agreement rate q* as the agreement rate predicted by quantum mechanics (Equation 3.1). The corresponding *disagreement rate* is $\widetilde{q} = 100 - q$.

Why Agreement Rates Involve $\cos^2 \gamma$

The presence of the square term in the quantum mechanical agreement rate, Equation 3.1, can be understood in terms of classical physics. Consider a filter C with a vertical polarising angle $\alpha_C = 0°$ and a light wave with a polarised angle β. In general, the intensity of a wave is proportional to the square of its amplitude; so if the light intensity is proportional to $\cos^2(\alpha - \beta)$ then there must be a wave with amplitude $A \propto \cos(\alpha - \beta)$. For simplicity, we assume that the wave has an amplitude of 1, which can be decomposed into two components, where one component is parallel to the filter polarising angle α and the other is orthogonal to α, as shown in Figure 3.4. Because α is vertical here, the amplitude of the wave component parallel to α lies along the vertical axis and is equal to $y = \cos|\alpha - \beta|$. Similarly, the amplitude of the wave component orthogonal to α lies along the horizontal axis and is equal to $z = \sin|\alpha - \beta|$.

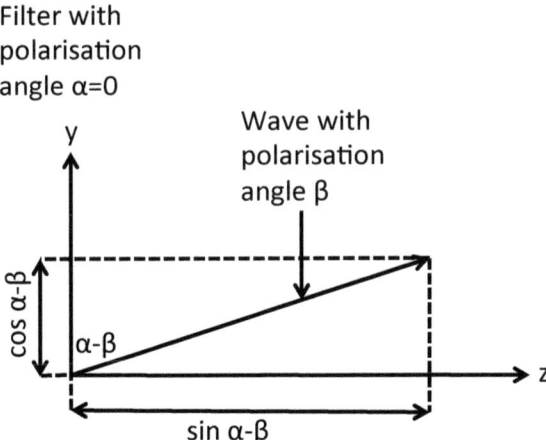

Figure 3.4: A wave with polarised angle β meets a filter with vertical polarising angle $\alpha = 0°$, which is parallel to the y-axis (so $\alpha - \beta$ is the difference between the wave polarised angle and the filter polarising angle). The wave (represented by the diagonal vector) has unit amplitude, which can be decomposed into a component $\cos(\alpha - \beta)$ that is parallel to the filter's polarising angle and a component $\sin(\alpha - \beta)$ that is orthogonal to the filter's polarising angle.

Because wave energy is proportional to the square of wave amplitude, the components of energy along the vertical and horizontal axes are

$$E_V = \cos^2(\alpha - \beta) \text{ J}, \qquad (3.2)$$

$$E_H = \sin^2(\alpha - \beta) \text{ J}, \qquad (3.3)$$

where $\cos^2(\alpha - \beta) + \sin^2(\alpha - \beta) = 1$. As Malus discovered in 1810, if a wave with polarised angle β meets a filter with vertical polarising angle α then a proportion $\cos^2(\alpha - \beta)$ of the wave's energy is transmitted by the filter, while the remainder $\sin^2(\alpha - \beta)$ is absorbed.

However, considering individual photons, it is not possible for a fraction of a photon's energy to pass through a filter, because it is not possible for a fraction of a photon to pass through a filter. Instead, each photon passes through a filter with a *probability* that is proportional to the square of the photon's wave amplitude along the filter's polarising angle. This means that each photon passes through the filter with a probability of $\cos^2(\alpha - \beta)$. Thus, the smaller the angular difference between the photon's polarised angle and the filter's polarising angle, the more likely it is that the photon will pass through the filter.

When considered over large numbers of photons, both the quantum mechanical account and the classical account predict the same amount of energy passing through the filter. Specifically, quantum mechanics predicts that a proportion $\cos^2(\alpha - \beta)$ of all photons will pass through the filter, whereas classical physics predicts that a proportion $\cos^2(\alpha - \beta)$ of the light's energy will pass through the filter. This is consistent with Bohr's *correspondence principle*, which states that the predictions of quantum mechanics should be consistent with classical physics (e.g. when large numbers of particles are involved).

Bizarre Agreement Rates

Consider three filters A, B and C, with polarising angles

$$\alpha_A = 0°, \qquad (3.4)$$
$$\alpha_B = +30°, \qquad (3.5)$$
$$\alpha_C = -30°, \qquad (3.6)$$

where $0°$ is vertical. If filters B and C are lined up in a row then the agreement rate is $q(B, C) = 25\%$, meaning that 25% of the photons that pass through B also pass through C, as shown in Figure 3.5a. But (as explained below) if the filter A is now inserted between B and C, then 56% of the photons that pass through B will pass through C, as shown in Figure 3.5b. In other words, inserting the filter A between B and C *increases* the percentage of photons that pass through C after they have passed through B, from $q(B, C) = 25\%$ to $q(B, A, C) = 56\%$. This is surprising because common sense tells us that placing a filter between B and C should reduce the amount of light that reaches C and therefore reduce the total amount of light that passes through C; after all, a filter should only be able to remove light, not add to it.

We can use Equation 3.1 to calculate how the agreement rates quoted above were obtained under quantum mechanics. The percentage of photons that pass through C after they have passed through B depends on the polarising angles of B and C. For the polarising angles in Equations 3.5 and 3.6, the difference is $\Delta\alpha_{BC} = 60°$, which implies an agreement rate of

$$
\begin{aligned}
q(B, C) &= 100 \times \cos^2(60°) \ \% \\
&= 25\%.
\end{aligned}
\tag{3.7}
$$

Similarly, the proportion of photons that pass through A after they have passed through B depends on the difference between the polarising

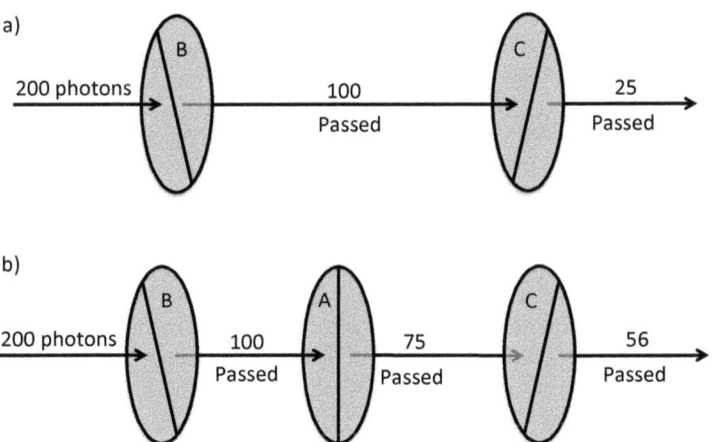

Figure 3.5: Adding a filter can increase the proportion of photons that pass through another filter. If 200 photons arrive at filter B with polarising angle $\alpha_B = -30°$ then 100 will pass through it and 100 will be absorbed. a) At filter C with $\alpha_C = +30°$, $q(B, C) = 25\%$ so 25 photons pass through. b) If a filter A with $\alpha_A = 0°$ is inserted between B and C, 75% of the 100 photons that passed through B will also pass through A, and $q(A, C) = 75\%$ of those will then pass through C, so that $q(B, A, C) = 0.75 \times 0.75 = 56\%$. Thus, adding A more than doubles the number of photons passed by C.

angles of B and A, which is $\Delta\alpha_{AB} = 30°$, so

$$
\begin{aligned}
q(B, A) &= 100 \times \cos^2(30) \% \\
&= 75\%.
\end{aligned}
\tag{3.8}
$$

Just as the difference $\Delta\alpha_{BA} = 30°$ allows 75% of the photons that passed through B to pass through A, so $\Delta\alpha_{AC} = 30°$ implies that

$$
\begin{aligned}
q(A, C) &= 100 \times \cos^2(30) \% \\
&= 75\%.
\end{aligned}
\tag{3.9}
$$

Therefore, the percentage of photons that pass through C after they have passed through both B and A is

$$
\begin{aligned}
q(B, A, C) &= 75\% \times 75\% \\
&\approx 56\%.
\end{aligned}
\tag{3.10}
$$

Thus, using the formula $q = \cos^2 \gamma$ (Equation 3.1), we can understand how inserting an additional filter (A) can increase the probability that a single photon will pass through another filter (C).

3.4. The Inverse Quantum Zeno Effect

The counter-intuitive trick of adding filters to increase the proportion of photons that pass through a destination filter can be used to force this proportion to approach 100%. For all practical purposes, a polarising filter acts as a measuring device that makes a *quantum measurement*. In other words, a filter with a polarising angle α forces the indeterminate quantum state of a photon to adopt one of two polarised angles. If a photon adopts the state $\beta = \alpha$ then it passes through the filter, but if it adopts the state $\beta = \alpha + 90°$ then it is absorbed by the filter. Alternatively, we could say that a photon which passes through the filter ends up with the state $\beta = \alpha$, but one that is absorbed by the filter ends up with the state $\beta = \alpha + 90°$ (which is odd because an absorbed photon has no state).

Consider a filter A with polarising angle $\alpha_A = 0°$. Next, add three filters, F_1, F_2 and F_3, with polarising angles $\alpha_1 = 30°$, $\alpha_2 = 60°$ and $\alpha_3 = 90°$. Because the difference between the polarising angles of A and F_1 is $\Delta\alpha = 30°$, $\cos^2(30°) = 75\%$ of photons that passed through A will pass through F_1. Similarly, the difference between the polarising angles of F_1 and F_2 is $\Delta\alpha = 30°$, so (again) 75% of photons that passed through F_1 will pass through F_2. Finally, the difference between the polarising angles of F_2 and F_3 is also $\Delta\alpha = 30°$, so 75% of photons that passed through F_2 will pass through F_3. Therefore, the percentage of photons that pass through filters F_1, F_2 and F_3, given that they passed through filter A, is (ignoring rounding errors)

$$q(A, F_1, F_2, F_3) = 0.75^3 \qquad (3.11)$$
$$= 42\%. \qquad (3.12)$$

But we know that in the absence of the filters F_1 and F_2, the percentage of photons that would pass through F_3 (which has a polarising angle of $\alpha_3 = 90°$) is $q(A, F_3) = 0\%$. So adding F_1 and F_2 increases that percentage from 0% to 42%. Can we do better? Yes.

If we add a series of 90 filters $F_1, \ldots, F_{90}$, where the difference between the polarising angles of consecutive filters is $\Delta\alpha = 1°$, then the percentage of photons that pass through all of $F_1, \ldots, F_{90}$, given that they passed through filter A, is

$$q(A, F_1, \ldots, F_{90}) = (\cos^2(1°))^{90} \qquad (3.13)$$
$$= 0.9997^{90} \qquad (3.14)$$
$$= 97.30\%. \qquad (3.15)$$

In general, if the polarising angle of the final filter F_n is $\alpha_n = 90°$ and we add a series of n filters $F_1, \ldots, F_n$, where the difference between the polarising angles of adjacent filters is $\Delta\alpha = 90/n$, then the percentage of photons that pass through filters $F_1, \ldots, F_n$, given that they passed through filter A, is

$$q(A, F_1, \ldots, F_n) = \cos^{2n}(\Delta\alpha). \qquad (3.16)$$

In the limit as $n \to \infty$, we have that $\Delta\alpha \to 0$ and therefore $q \to 100\%$. Thus, given a sufficiently large number of quantum measurement devices (filters) between the filters A and F_n, it is possible to coerce almost 100% of the photons to pass through the final filter F_n. This is remarkable, because in the absence of those quantum measurement devices, $q(A, F_n) = 0\%$ of the photons would pass through F_n.

This phenomenon, in which the act of measuring a photon at a contiguous series of polarising angles effectively forces the photon to adopt a corresponding series of states, is called the *inverse quantum Zeno effect*. Here, a state corresponds to a particular polarised angle, but it can refer to any physical quantity, such as position or momentum.

These results seem plausible, or at least possible, for a single photon because it is the same photon that tries to pass through the filters. However, *it seems much less plausible that quantum mechanics could explain how one photon passing through filter B can affect the probability that* a different photon *will pass through filter C*.

The following three sections provide different accounts of why classical physics cannot, in principle, account for the strange behaviour of pairs of photons. Section 3.5 gives a brief self-contained account. Section 3.6 is based on a statistical method due to Guy Blaylock, whereas Section 3.7 is based on David Mermin's counting method.

3.5. The Short Version

Filters that polarise light are now commonplace, and are often used in sunglasses. If the lenses from a pair of polarising sunglasses are removed, they can be used to measure the polarisation angle of individual photons. Consider an experiment in which two people, Alice and Bob, are a mile apart. Every second, a pair of entangled photons is created, and one heads towards Bob while the other heads towards Alice. For simplicity, we assume that both photons adopt the same polarisation angle. Incidentally, creating a pair of entangled photons is a fairly technical process, beyond the scope of this book.

To measure photons, Alice and Bob use filters whose orientations can be adjusted to vary the agreement between them. For example, if the filters are rotated to have the same polarising angle then for any photon that passes through Bob's filter its corresponding one (of the entangled pair) also passes through Alice's filter, and for any photon absorbed by Bob's filter the corresponding one is also absorbed by Alice's filter. If only Bob rotates his filter by an angle $\Delta\alpha_B = 30°$, they find that the disagreement rate rises from 0% to 25%. In this case, rotating Bob's filter means that 25% of the photons that would have behaved like Alice's photons changed their behaviour. Similarly, if only Alice rotates her filter in the opposite direction, $\Delta\alpha_A = -30°$, then the disagreement rate also rises to 25%.

Now suppose Bob rotates his filter by $\Delta\alpha_B$ and Alice rotates her filter by $\Delta\alpha_A$. In this case, 25% of Bob's photons that would have behaved like Alice's photons change their behaviour, and 25% of Alice's photons that would have behaved like Bob's photons change their behaviour, so the disagreement rate should be no greater than $25\% + 25\% = 50\%$. However, if this experiment were to be carried out, we would find that the disagreement rate is not 50% but 75%. Clearly, this result defies the common sense prediction of 50%, but it is consistent with the predictions of quantum mechanics.

3.6. Counting Photon Pairs: Bell's Inequality

Consider pairs of *entangled* photons P_1 and P_2. In general, P_1 and P_2 can have any polarisation angles, but to keep matters simple we assume that they have equal polarisation angles. Now send P_1 towards filter A and send P_2 towards filter B, where A and B lie in opposite directions. As observed in experiments, an average of half of all photons in unpolarised light will pass through any filter, irrespective of its polarising angle. To comply with this condition, an average of half of all P_1 photons will pass through filter A irrespective of its polarising angle, and half of all P_2 photons will pass through filter B irrespective of its polarising angle. The polarising angles of A and B are

$$\alpha_A = 0°, \qquad (3.17)$$
$$\alpha_B = 30°, \qquad (3.18)$$

as in Figure 3.6. Next — wait — A is on Mars and B is on Pluto.

We define the *agreement rate for photon pairs* as the percentage of pairs of photons (P_1, P_2) for which P_1 passes through its filter and P_2 passes through its filter; this is also the proportion of pairs of photons such that P_1 is absorbed by its filter and P_2 is absorbed by its filter. If the percentage of pairs of photons that agree is $q(A, B) = 75\%$ then the percentage of pairs of photons in which P_1 passes and P_2 is absorbed, or *vice versa*, yields a *disagreement rate* of $\tilde{q}(A, B) = 25\%$. By analogy with Equation 3.1, quantum mechanics predicts that the percentage of these photon pairs that will pass through both of filters A and B is

$$q(A, B) = (100 \times \cos^2 \gamma) \% \qquad (3.19)$$

where $\gamma = \alpha_A - \alpha_B$, which has the same form as in Figure 3.3.

When measured over many repeated experiments, Equation 3.19 correctly predicts that if P_1 passes through A then 75% $(= \cos^2(-30°))$ of P_2 photons pass through B, and if P_1 is absorbed by A then 75% of P_2 photons are absorbed by B. In other words, 75% of all photon pairs somehow agree that they will both pass through or both be absorbed by their respective filters, as in Figure 3.6.

As was the case for the single-photon experiments, the agreement rate depends on the exact nature of the filters used. To take an extreme example, if the polarising angle of B is set to α_A (so that the difference in polarising angles of the filters is 0°) then $q(A, B)$ instantly jumps to 100%. In contrast, if the polarising angle of B is set to $\alpha_B = 90°$ then $q(A, B)$ instantly falls to 0%. Somehow, the polarising angle of the Pluto filter changes the proportion of photons that pass through the Mars filter.

We can translate the above description into a tangible example consisting of eight experiments. Note that the results below are not meant to be taken as the results of eight actual experiments, but rather are intended to represent the overall statistical structure of a long series of experiments. An agreement rate of $q(A, B) = 75\%$ is shown here for a set of eight pairs of photons labelled a–h:

$$
\begin{array}{lcc}
\text{Pair} & \text{a b c d} & \text{e f g h} \\
P_1 \rightarrow \text{Filter A} & \text{1 1 1 1} & \text{0 0 0 0} \\
P_2 \rightarrow \text{Filter B} & \text{0 1 1 1} & \text{0 0 0 1}
\end{array}
\tag{3.20}
$$

where each column represents the outcome of a single experiment with an entangled pair of photons. The row labelled Filter A shows the outcomes of the P_1 photons of the eight pairs, and the row labelled Filter B shows the outcomes of the P_2 photons of those pairs. The binary digit 0 indicates that the photon is absorbed by the filter named on the left, and 1 indicates that the photon passes through that filter. For the four pairs a–d, photon P_1 always passes through its filter (A), whereas photon P_2 passes through its filter (B) only for pairs b, c and d. For the four pairs e–h, photon P_1 is always absorbed by its filter (A), whereas photon P_2 is absorbed by its filter (B) only for pairs e, f and g. Only the middle six pairs have matching behaviours, so the agreement rate is $q(A, B) = 6/8 = 75\%$. Notice that 4 out of 8 photons

a)

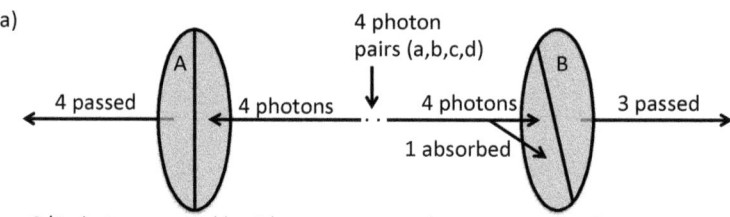

3/4 photons passed by A have a partner photon passed by B.

b)

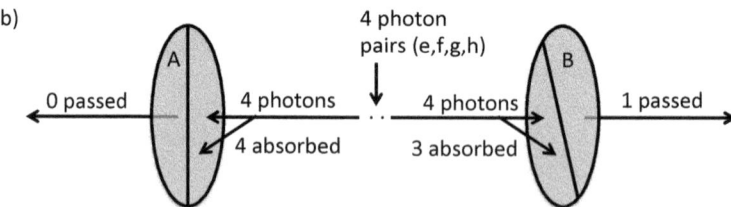

3/4 photons absorbed by A have a partner photon absorbed by B.

Figure 3.6: Bell's inequality for filters with $\alpha_A = 0°$ and $\alpha_B = 30°$. In each of eight pairs a–h of entangled photons, photon P_1 travels to filter A and photon P_2 travels to filter B. For these filters, there is a 75% agreement rate:
a) For pairs a–d, four P_1 photons pass through A and three of their partner photons P_2 pass through B.
b) For pairs e–h, four P_1 photons are absorbed by A and three of their partner photons P_2 are absorbed by B. All figures are averages.

pass through each filter, in accordance with what is known about the average proportion of photons passing through any filter.

At this point, we define the *common-sense agreement rate* Q as the agreement rate predicted by common sense, as explained below. The corresponding *disagreement rate* is $\widetilde{Q} = 100 - Q$.

Next, we derive a common-sense agreement rate $Q(A, C)$ for filters A and C, where C has a polarising angle $\alpha_C = -30°$. In this case, $\Delta\alpha = |\alpha_A - \alpha_C| = 30°$, as was the case for filters A and B. Given that the magnitude of $\Delta\alpha$ is unchanged upon replacing B (with $\alpha_B = 30°$) by C (with $\alpha_C = -30°$), the common-sense disagreement rate $\widetilde{Q}$ is also unchanged. For example, in the set of eight pairs of photons for filters A and C, only the pairs d and e disagree:

$$
\begin{array}{llll}
\text{Pair} & \text{a b c d} & \text{e f g h} \\
P_1 \to \text{Filter A} & \text{1 1 1 1} & \text{0 0 0 0} \\
P_2 \to \text{Filter C} & \text{1 1 1 0} & \text{1 0 0 0}
\end{array}
\qquad (3.21)
$$

so that $\widetilde{Q}(A, C) = \widetilde{Q}(A, B) = 25\%$.

Now suppose we compare a typical sequence of outcomes for filter B with a typical sequence of outcomes for filter C. If the photons are extremely lucky then the rows of outcomes for B and C are identical:

$$
\begin{array}{llll}
\text{Pair} & \text{a b c d} & \text{e f g h} \\
P_1 \to \text{Filter B} & \text{0 1 1 1} & \text{0 0 0 1} \\
P_2 \to \text{Filter C} & \text{0 1 1 1} & \text{0 0 0 1}
\end{array}
\qquad (3.22)
$$

Therefore, common sense indicates that the smallest possible disagreement rate is $\widetilde{Q}(B, C) = 0\%$. At the other extreme, the outcomes for B could be the same as in Equation 3.20 and the outcomes for C could be the same as in Equation 3.21:

$$
\begin{array}{llll}
\text{Pair} & \text{a b c d} & \text{e f g h} \\
P_1 \to \text{Filter B} & \text{0 1 1 1} & \text{0 0 0 1} \\
P_2 \to \text{Filter C} & \text{1 1 1 0} & \text{1 0 0 0}
\end{array}
\qquad (3.23)
$$

It is important to note that *this represents the maximum possible disagreement rate for filters B and C, given the previously observed disagreement rates between A and B and between A and C.* In other words, if the disagreement rate for filters A and B is 25% and the disagreement rate for filters A and C is also 25% then common sense dictates that the disagreement rate for filters B and C cannot be greater than $25\% + 25\% = 50\%$. Given the two extreme scenarios represented in Equations 3.22 and 3.23, the predicted common-sense disagreement

3 Einstein's Unreasonable Reality

rate should be between 0% and 50%,

$$\widetilde{Q}(\mathrm{B,C}) \ \leq \ 50\%. \qquad (3.24)$$

This is *a form of Bell's inequality*.

However, the quantum mechanical disagreement rate for B and C is substantially greater than this common-sense prediction of 50%. Specifically, as $\Delta\alpha = 60°$ for filters B and C, quantum mechanics predicts an agreement rate of $q(\mathrm{B,C}) = \cos^2(60°) = 25\%$, which implies a disagreement rate of

$$\widetilde{q}(\mathrm{B,C}) \ = \ 75\%. \qquad (3.25)$$

An observed disagreement rate of more than 50% would violate Bell's inequality (Equation 3.24). In practice, the disagreement rate for filters B and C is 75%, as correctly predicted by quantum mechanics.

3.7. The Hidden-Variable Hypothesis

As was mentioned in Section 3.1, there is one explanation for the type of behaviour described above that sounds reasonable but which is provably wrong. If both P_1 and P_2 carry a copy of the same computer program, so that both photons in each pair behave in the same way when they meet the same type of filter, then they must always *agree* on whether or not to pass through a given type of filter. For example, a program might specify that photons pass through filters A and B but not C. Because this assumes that both photons have access to the same program, which contains variables that are hidden from any observer, it is a type of hidden-variable hypothesis.

What is unreasonable, and which defies common sense, is that if the hidden-variable hypothesis were true then it would be impossible for photons to match the behaviour predicted by quantum mechanics.

The precise values of the polarising angles of filters will not matter until later, but we use specific values here so that we can proceed with tangible examples. Consider filters A, B and C with polarising angles

$$\alpha_\mathrm{A} = 0°, \quad \alpha_\mathrm{B} = +60°, \quad \alpha_\mathrm{C} = -60°, \qquad (3.26)$$

and define $\Delta\alpha = 60°$. For these filters, the difference between each pair is tabulated in Figure 3.7.

We will consider many combinations of photons and filters, so we need notation to specify each combination. As a defining example, suppose that the program given to a photon is a triplet of instructions $\mathrm{T} = [\mathrm{A}{=}1, \mathrm{B}{=}1, \mathrm{C}{=}0]$, where A=1 means that the photon passes through filter A, B=1 means that the photon passes through filter B, and

C=0 means that the photon is absorbed by filter C. The program T $= [A=1, B=1, C=0]$ can be abbreviated to T $= [110]$.

Given that a photon either passes through (1) or is absorbed (0), there are two possible outcomes for each photon (0 or 1), and because there are three filters (A, B and C), this means there are $2^3 = 8$ possible programs:

$$T1 = [000], \quad T2 = [001], \quad T3 = [010], \quad T4 = [011],$$
$$T5 = [100], \quad T6 = [101], \quad T7 = [110], \quad T8 = [111]. \tag{3.27}$$

Now, consider what happens if we attach copies of the same program to P_1 and P_2 and then choose a filter for each of P_1 and P_2 at random. For example, in the first experiment we attach a copy of T7 to both photons, which specifies that they pass through filters A and B but not C.

Using the program T7 $= [110]$, if the filters chosen at random are A for P_1 and B for P_2 then both photons pass through their respective filters (represented as 11). In contrast, if the chosen filters are A for P_1 and C for P_2 then P_1 passes through its filter but P_2 does not. This result can be represented as 10 or 01; the order of the binary digits is immaterial here, because we are interested only in the agreement rate of the photons. Note that there are exactly $3^2 = 9$ pairs of filter combinations, so we can make a *truth table* that specifies the behaviour of P_1 and P_2 when confronted with each of these nine possible pairs of filters, as in Figure 3.8a. Over the set of nine possible pairs of filters, we find that the two photons agree (i.e. both do or both don't pass through their respective filters) in 5 out of the 9 cases for the program T7, as shown in Figure 3.8b. This implies a *hidden-variable agreement rate* for T7 of

$$Q_7 = 5/9 \approx 56\%. \tag{3.28}$$

Over a long run of experiments, all pairs of filters will be chosen equally often, so the agreement rate $Q_7 = 56\%$ represents an average for T7.

Filter	A	B	C
A	0°	60°	60°
B	60°	0°	120°
C	60°	120°	0°

(a)

Filter	A	B	C
A	0	$\Delta\alpha$	$\Delta\alpha$
B	$\Delta\alpha$	0	$2\Delta\alpha$
C	$\Delta\alpha$	$2\Delta\alpha$	0

(b)

Figure 3.7: (a) Absolute value of the difference in polarising angle α for each pair of filters (Eq. 3.26). (b) Values expressed in units of $\Delta\alpha$.

It turns out that the six programs T2–T7 all yield the same agreement rate (56%). For each of these programs, the photons in each pair agree in five out of nine choices of filter pairs. This is because each of these programs contains either two 1s and one 0, or two 0s and one 1, as shown in Figure 3.9.

In contrast, the program T1 contains three 0s, so every choice of the nine possible filter pairs yields agreement. Similarly, program T8 contains three 1s, so (again) every choice of the nine possible filter pairs yields agreement. Therefore, programs T1 and T8 have $Q_1 = Q_8 = 9/9 = 100\%$.

When considered over the eight programs T1–T8, the hidden variable prediction of the average percentage of photon pairs in which both photons agree is

$$\overline{Q} = (Q_1 + Q_2 + Q_3 + Q_4 + Q_5 + Q_6 + Q_7 + Q_8)/8. \qquad (3.29)$$

The agreement rate for each of the programs T2–T7 is 5/9, and agreement rate for T1 and T8 is 9/9, so the average agreement predicted by the hidden-variable hypothesis is

$$\begin{aligned} \overline{Q} &= (9/9 + 5/9 + 5/9 + 5/9 + 5/9 + 5/9 + 5/9 + 9/9)/8 \\ &\approx 67\%. \end{aligned} \qquad (3.30)$$

Strictly speaking, this is not a form of Bell's inequality, but only because it specifies a value (rather than a bound); nevertheless, it provides a

	A	B	C		A	B	C
A	11	11	10	A	Agree	Agree	**Disagree**
B	11	11	10	B	Agree	Agree	**Disagree**
C	10	10	00	C	**Disagree**	**Disagree**	Agree
	(a)				(b)		

Figure 3.8: Behaviour of a pair of photons P_1 and P_2 on encountering a pair of filters. Each photon has a copy of program T7 = [110], which specifies that it should pass through filters A and B but not C.
(a) Each cell indicates how each photon in a pair behaves for one choice of filter pairs. For example, 10 (top right) means that photon P_1 passes through its filter (A) but its partner photon P_2 does not (i.e. it is absorbed by filter C).
(b) Behaviour of photon pairs expressed as Agree or Disagree for program T7. If both photons in a pair pass (11), or if both photons are absorbed (00), then they Agree; else they Disagree (i.e. 01 or 10). For programs T2–T7 (Equation 3.27), photon pairs agree in 5 out of 9 cases. For T1 and T8 they agree in 9 out of 9 cases.

test of the hidden-variable hypothesis. Specifically, if a series of physical experiments identical to those above yields an average agreement rate $\bar{q}$ that differs from $\overline{Q} = 67\%$ then the hidden-variable hypothesis is false.

Note that, when implemented as described above, the hidden-variable hypothesis predicts that $\overline{Q}$ is constant and so its value is independent of the polarising angles of the three filters A, B and C. This alone gives a strong hint that the hidden-variable hypothesis cannot be true.

The average agreement rate $\bar{q}$ obtained in physics experiments, as correctly predicted by quantum mechanics, is calculated as follows. As a reminder, given two filters with polarising angles α_A and α_B for which the absolute value of the difference is $\gamma = |\alpha_A - \alpha_B|$, the observed

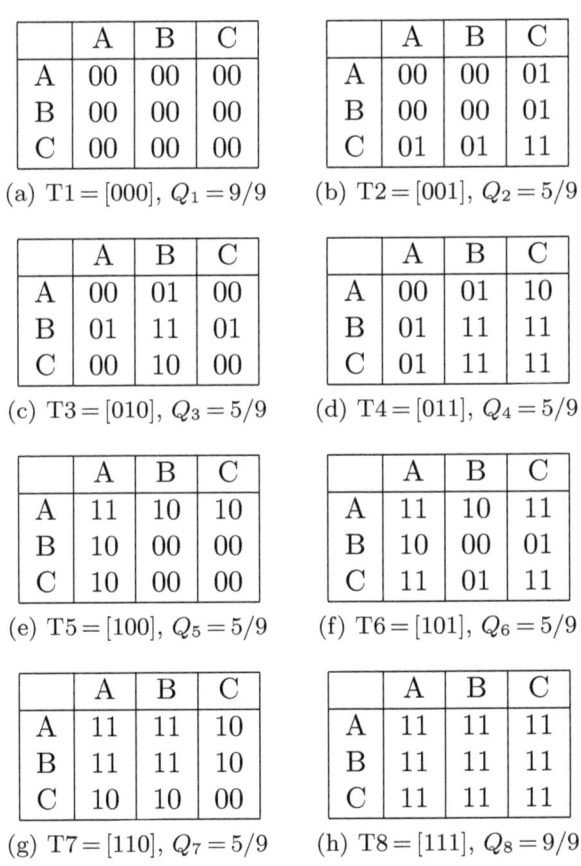

	A	B	C
A	00	00	00
B	00	00	00
C	00	00	00

(a) T1 = [000], $Q_1 = 9/9$

	A	B	C
A	00	00	01
B	00	00	01
C	01	01	11

(b) T2 = [001], $Q_2 = 5/9$

	A	B	C
A	00	01	00
B	01	11	01
C	00	10	00

(c) T3 = [010], $Q_3 = 5/9$

	A	B	C
A	00	01	10
B	01	11	11
C	01	11	11

(d) T4 = [011], $Q_4 = 5/9$

	A	B	C
A	11	10	10
B	10	00	00
C	10	00	00

(e) T5 = [100], $Q_5 = 5/9$

	A	B	C
A	11	10	11
B	10	00	01
C	11	01	11

(f) T6 = [101], $Q_6 = 5/9$

	A	B	C
A	11	11	10
B	11	11	10
C	10	10	00

(g) T7 = [110], $Q_7 = 5/9$

	A	B	C
A	11	11	11
B	11	11	11
C	11	11	11

(h) T8 = [111], $Q_8 = 9/9$

Figure 3.9: Behaviour of a pair of photons P_1 and P_2 on encountering each of three pairs of filters (AB, AC and BC), for each of eight possible programs T1–T8. Both photons in each pair carry a copy of the same program Ti, and the agreement rate Q_i for each program is expressed as a fraction. Averaged over all eight programs, $\overline{Q} = 67\%$ (Eq. 3.30).

percentage of entangled photon pairs in which both photons agree is $q_{\text{pair}} = \cos^2 \gamma$. When considered over the nine possible filter pairs (Figure 3.7b), we find that

$$73 \text{ filter pairs yield } \gamma = 0°,$$
$$4 \text{ filter pairs yield } \gamma = \Delta\alpha,$$
$$2 \text{ filter pairs yield } \gamma = 2\Delta\alpha.$$

If the filters A, B and C are chosen at random over a long series of experiments then each of the nine possible filter pairs would be chosen equally often. Therefore, the quantum mechanical prediction of the average percentage of entangled photon pairs in which both photons agree is

$$\bar{q}(\Delta\alpha) = (3\cos^2 0 + 4\cos^2 \Delta\alpha + 2\cos^2 2\Delta\alpha)/9. \qquad (3.31)$$

If we set $\alpha = 60°$ then $\Delta\alpha = 60°$, $\cos\Delta\alpha = 0.5$ and $\cos 2\Delta\alpha = -0.5$, so

$$\begin{aligned} \bar{q}(60°) &= ((3 \times 1) + (4 \times 0.5^2) + (2 \times (-0.5)^2))/9 \\ &= 50\%. \end{aligned} \qquad (3.32)$$

The value $\Delta\alpha = 60°$ was chosen specifically to yield $\bar{q} = 50\%$. The difference between the hidden-variable hypothesis value of $\overline{Q} = 67\%$ and

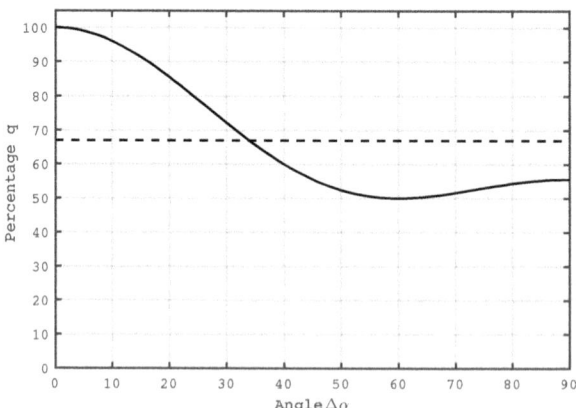

Figure 3.10: Three filters A, B and C have polarising angles $\alpha_A = 0$, $\alpha_B = +\alpha$ and $\alpha_C = -\alpha$, with $\Delta\alpha = |\alpha_C - \alpha_A| = |\alpha_B - \alpha_A|$. If two filters are chosen equally often (e.g. at random) from A, B and C then the programs T1–T8 predict a constant average agreement rate of $\overline{Q} \approx 67\%$ (dashed line, Equation 3.30). In contrast, quantum mechanics predicts an agreement rate $\bar{q}$ that varies with $\Delta\alpha$ (solid curve, Equation 3.31).

the quantum mechanical value of $\overline{q} = 50\%$ may not seem like much, but it is an unbreachable barrier to any hidden-variable model. A complete record of $\overline{Q}$ (as predicted by the hidden-variable hypothesis, Equation 3.30) and $\overline{q}$ (as predicted by quantum mechanics, Equation 3.31) for $0° < \Delta\alpha < 90°$ is displayed in Figure 3.10. This shows that the single $\overline{Q}$ value predicted by the hidden-variable hypothesis and the $\overline{q}$ values predicted by quantum mechanics are different at every value of $\Delta\alpha$ except $\Delta\alpha \approx 34°$.

In summary, if the photons P_1 and P_2 each carry a copy of the same program, which guarantees that they would agree (i.e. both pass through or both be absorbed) when confronted with the same filter, and if the filter for each photon is chosen at random from A, B and C, then $\overline{Q} = 67\%$ of photon pairs would agree.

Crucially, if any triplet of filters exists such that physical experiments yield an average agreement rate $\overline{q}$ that is not equal to 67% then this would rule out the hidden-variable hypothesis. For the filters A, B and C above with $\Delta\alpha = 60°$, the average agreement rate given by quantum mechanics is 50%, which is incompatible with the hidden-variable hypothesis.

Finally, disproving the hidden-variable hypothesis has two vital implications. First, it rules out the possibility that both photons in an entangled pair have the same polarised angle before being measured. If the photons had a definite value of polarised angle then the hidden-variable hypothesis could not be disproved, which brings us to the second, and more important, implication. Disproving the hidden-variable hypothesis rules out the possibility that each photon has *any* polarised angle before it is measured. In other words, the polarised angle of each photon is *undefined* before it is measured. This conclusion applies not only to the polarisation states of photons but also to any observable physical quantity, such as the position or momentum of photons, electrons, atoms, molecules and (perhaps) cats.

Faster-Than-Light Communication?

Consider two entangled photons P_1 and P_2 with polarisation angles measured when the photons are located at x_1 and x_2, respectively. If the delay between measurements made at x_1 and x_2 is long enough to allow light (and therefore information) to pass from x_1 to x_2 then it is possible that the observed correlation between polarisation angles is due to communication between x_1 and x_2. This is called the *locality loophole*, which has been closed[14] by ensuring that the delay between measurements is shorter than the time it takes light to pass from x_1 to x_2. To make use of this faster-than-light communication, we would have to be able to control the polarising angle of each photon being measured. But a fundamental aspect of quantum systems is that the measured polarising angle is random, which effectively precludes any observer from using entangled pairs to send or receive any information faster than the speed of light.

Entanglement and Decoherence

If two particles are entangled then any form of disturbance seems to act as a measurement, which forces the particles to behave like classical (i.e. disentangled) particles. For example, we shall see in the next chapter how measuring light waves forces them to materialise into photons, which behave like classical particles. Indeed, even the normal thermal fluctuations of atoms may be sufficient to cause entangled particles to become disentangled. For all practical purposes, entangled particles are effectively measured all the time by the random fluctuations in their local environment. Disentangling particles is called *decoherence*[34].

3.8. Summary

Einstein did not think that quantum mechanics is wrong, but he did think it is incomplete. For several decades, the EPR paper (published in 1935) represented a challenge to the foundations of quantum mechanics because it implied that entangled particles could affect each other without any identifiable force or means of communication. The problem was that no-one knew how to test the outrageous predictions of the EPR paper. Then, in 1964, John Bell produced his (now famous) theorem, which finally showed how the EPR predictions could be tested experimentally. After a relatively short interval, the first experimental results seemed to suggest a rare occurrence: Einstein was wrong (to believe that physically separate particles could not communicate with each other). Just to be clear, these experimental results confirmed the predictions of the EPR paper that physically separate entangled particles do seem to communicate with each other. Since that time, increasingly precise experiments have repeatedly confirmed the existence of non-locality (i.e. action at a distance) for entangled particles.

The importance of this result cannot be overstated. The EPR predictions and Bell's inequality represent a distillation of the most fundamental truths in quantum mechanics. The result of this distillation was aptly summarised by Richard Feynman:

> *I've entertained myself always by squeezing the difficulty*
> *of quantum mechanics into a smaller and smaller place, so*
> *as to get more and more worried about this particular item.*
> *It seems to be almost ridiculous that you can squeeze it to*
> *a numerical question that one thing is bigger than another.*
> *But there you are.*

Chapter 4

Waves of Light and Matter

I was guided by the aim to perform a real physical synthesis, valid for all particles, of the coexistence of the wave and of the corpuscular aspects that Einstein had introduced for photons in his theory of light quanta in 1905.
De Broglie L, 1970.

4.1. Introduction

In retrospect, it seems reasonable that an ethereal substance such as light can behave like a wave, but it seems less reasonable, and even ridiculous, to imagine that solid objects, from electrons to whole molecules, can behave like waves. In 1924, Prince Louis de Broglie proposed just that[7]. It was an extraordinary claim, but within four years de Broglie had been proved right.

4.2. Matter Waves

De Broglie proposed that a particle with momentum p has a wavelength

$$\lambda \ = \ h/p \ \text{m}, \tag{4.1}$$

where h is Planck's constant, as shown in Figure 4.1.

At first, this seems odd because momentum in classical physics is mass × velocity, but particles such as photons have no mass. However, according to Einstein's theory of special relativity (1905), photons have a *relativistic mass* M. Given that a photon travels at the speed of light $c \approx 3 \times 10^8$ m/s, its momentum is $p = Mc$. If we substitute this into Einstein's famous equation $E = Mc^2$ then we obtain

$$E \ = \ pc. \tag{4.2}$$

From Planck, we also know that

$$E \ = \ h\nu \ = \ hc/\lambda. \tag{4.3}$$

49

The equivalence of energy in Equations 4.2 and 4.3 yields

$$p \;=\; h/\lambda, \tag{4.4}$$

and therefore $\lambda = h/p$, as proposed by de Broglie.

The radical part of de Broglie's proposal was that the formula $\lambda = h/p$ should also apply to particles with nonzero mass. For example, an electron with mass $M = 9.109 \times 10^{-31}$ kg moving at a velocity $v \lll c$ has momentum $p = Mv$. (The proviso $v \lll c$ ensures that we can ignore relativistic effects associated with velocities close to c.) Therefore, de Broglie's equation implies a wavelength of $\lambda = h/(Mv)$, so wavelength decreases nonlinearly as momentum increases (Figure 4.1).

Confirmation that de Broglie's equation applies to particles with mass was published in 1927, when Davisson and Germer showed that electrons produce a diffraction pattern when they are reflected by crystalline nickel. In the same year, Thomson and Reid discovered that electrons produce interference patterns after passing through a thin film of celluloid.

If electrons are used in a double-slit experiment then the resultant interference pattern is qualitatively indistinguishable from the pattern obtained with light, as shown by Mollenstedt and Duker[25] in 1955 (except that the smaller wavelengths of electrons produces a pattern that is about 500 times smaller). Similarly, if the intensity of electrons is reduced until only one electron hits the screen at a time, an interference pattern gradually emerges[36], as if each individual electron is a wave (Figure 4.2). Evidence for the wave-like behaviour of electrons is now so

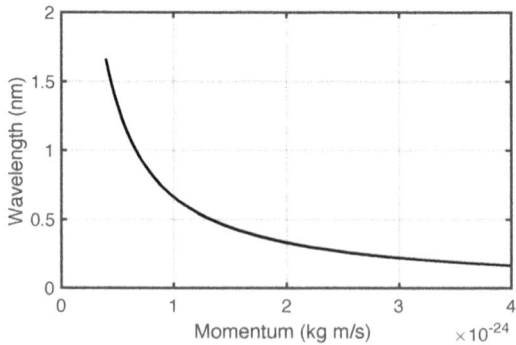

Figure 4.1: The relationship between momentum and de Broglie wavelength for an electron travelling at speeds between 0.44×10^6 m/s and 4.4×10^6 m/s. An electron with mass $M = 9.109 \times 10^{-31}$ kg orbits a hydrogen atom in its ground state at $v = 2.19 \times 10^6$ m/s, which implies a momentum of $p = Mv \approx 2 \times 10^{-24}$ and a wavelength of $\lambda = h/p \approx 0.33$ nm. For comparison, ultraviolet light has $\lambda \approx 400$ nm.

overwhelming that the waves associated with solid objects are known as *matter waves*. Even more remarkably, it has been demonstrated that whole atoms, and even complex molecules, are matter waves.

Note that de Broglie's insight means that all references to particles and waves are provisional. Sometimes it suits our purposes to speak of particles, but at other times it is more appropriate to speak of waves. This is the famous *wave–particle duality* (see Section 5.8). Crucially, because a wave is extended over space and time (e.g. Figure 5.6), the position and momentum of a particle associated with that wave are *undefined* until a measurement is made, whence their values have a precision that is limited by Heisenberg's uncertainty principle.

4.3. Heisenberg's Uncertainty Principle

Heisenberg's Ungenauigkeit or *uncertainty principle* states that for certain pairs of physical quantities, it is not possible to know the values of both members of the pair with absolute certainty[16]. Such pairs of quantities are called *complementary* or (more formally) *conjugate variables*, and include position and momentum (described below), as well as energy and time.

As stated in Chapter 1, the uncertainty principle was originally formulated in 1927 in German as *Ungenauigkeit*, which means inexactness or vagueness rather than uncertainty. Heisenberg proved that we cannot know both position and momentum exactly not just because we cannot measure them precisely, but because even in principle they cannot both be known exactly.

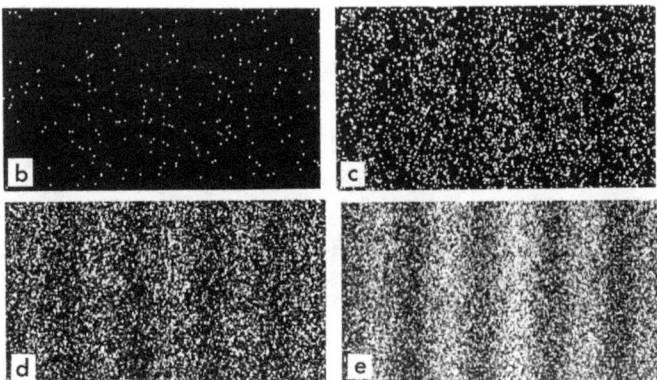

Figure 4.2: The gradual emergence of an interference pattern in a double-slit experiment[36]. Each dot represents the position of one electron, and the temporal sequence is b, c, d and then e. Reproduced with permission from Dr Tonomura's Wikimedia image.

Heisenberg's uncertainty principle permeates all of quantum mechanics, so it is important to have at least one convincing demonstration of its existence. The simplest demonstration involves diffraction at a single slit, which will also be useful when we come to consider the double-slit experiment in Chapter 5. However, it should be emphasized that these demonstrations are consequences, rather than causes, of Heisenberg's uncertainty principle, which is underpinned by formal proofs (see Heisenberg's inequality on p58).

4.4. Diffraction

The simplest form of camera forms an image through a pinhole aperture, as in Figure 4.3a. The optical characteristics of an imaging system are defined in terms of its ability to resolve small points of light. For example, a distant point of light, such as a star, acts as a *point source*, and the wave nature of light means that light diffracts around the aperture, producing an image that is a *Fraunhofer diffraction pattern*. This diffraction pattern has a central bright region (known as an *Airy*

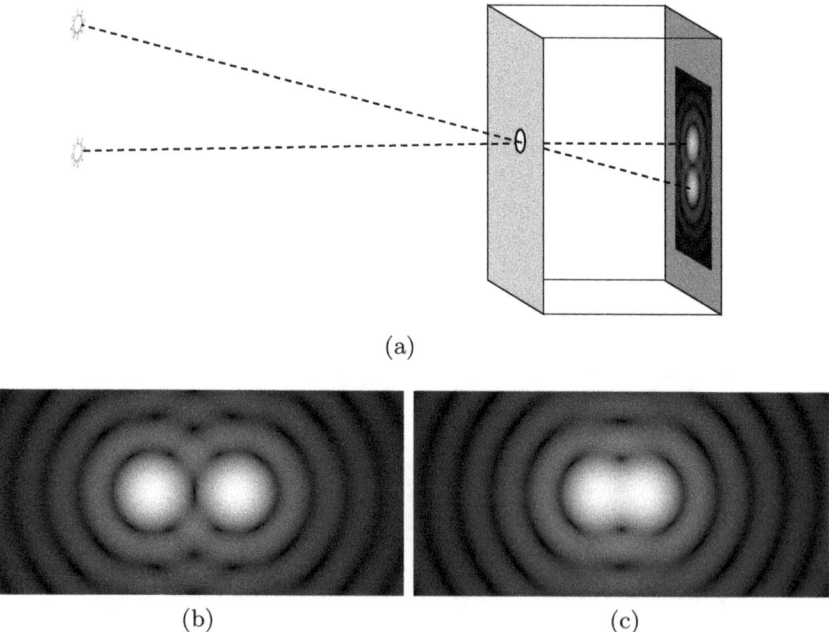

(a)

(b)   (c)

Figure 4.3: (a) Using a pinhole camera, point light sources at large distances form Airy discs. (b) Airy discs that can be resolved. (c) If the aperture is made smaller or if the wavelength is made longer then the Airy discs merge (Equation 4.9). Images by Spencer Bliven, reproduced with permission.

disc for circular apertures) surrounded by smaller side lobes, as shown in Figure 4.3b,c. The angular half-width of each bright region is the *resolution* of the imaging system, which determines how much detail can be seen. Thus, for a distant point of light, the diffraction pattern is an image of that point source. Clearly, if the images (diffraction patterns) of two point sources overlap then they cannot be differentiated in the image (e.g. Figure 4.3c).

If we consider a vertical cross-section through the pinhole aperture then the hole becomes a slit, as in Figure 4.4. A reasonably intuitive way to understand how the diffraction pattern forms is to assume that each point within the slit is the source of a new wave, called a *wavelet* (see Figure 4.4). Consider two points s_1 and s_2 within the slit, where one point, s_1, is at the top of the slit and the other, s_2, is at the centre of the slit, so that

$$\begin{aligned} \Delta s &= |s_2 - s_1| & (4.5) \\ &= d/2, & (4.6) \end{aligned}$$

where d is the width of the slit. We can assume that the paths from s_1 and s_2 to a point y on the screen are parallel to each other, because the distance D from the slit to the screen is large relative to Δs.

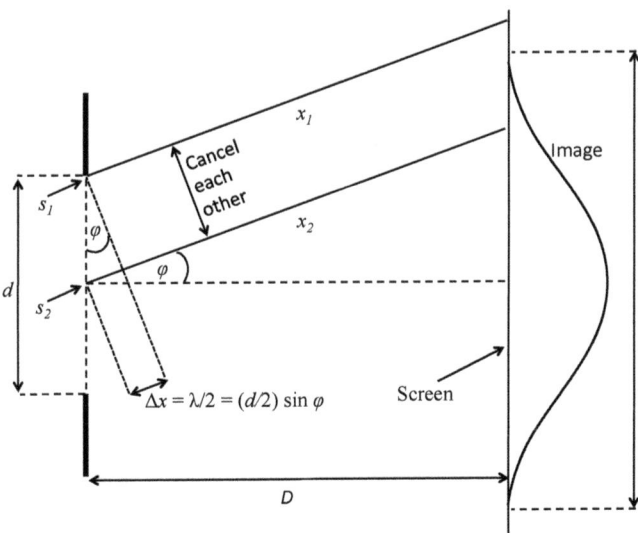

Figure 4.4: The central bright region of a diffraction pattern from a slit of width d, where the distance from slit to screen is D. If the ratio D/d is large then the lines labelled x_1 and x_2 are approximately parallel. The resolution of the system is half the angular width of the central bright region, which is φ radians where $\sin \varphi = \lambda/d$.

Consider a point y on the screen that lies in the direction φ. Compared to the wavelet from s_1, the wavelet from s_2 has to travel an extra distance Δx to reach y. From Figure 4.4 we see that

$$\sin \varphi \approx \Delta x / \Delta s, \tag{4.7}$$

where the approximation symbol $\approx$ is used because the lines labelled x_1 and x_2 are almost, but not exactly, parallel. As explained below (Equation 4.9), if the wavelength λ is much smaller than the slit width d then the 'amount' of diffraction is small and so the angle φ will be small. In this case we can use the small-angle approximation $\sin \varphi \approx \varphi$, so that Equation 4.7 becomes

$$\varphi \approx \Delta x / \Delta s \text{ radians.} \tag{4.8}$$

If the wavelets from s_1 and s_2 have the same phase at the point y on the screen then they reinforce each other, as in Figure 4.5(1). Conversely, if they have different phases at y then they cancel each other out, either partially as in Figure 4.5(2) and (3), or completely as in Figure 4.5(4). Now suppose that the point y is such that the extra path length from s_2 to it is exactly half a wavelength (i.e. $\Delta x = \lambda/2$); then the wavelets from s_1 and s_2 are in anti-phase when they reach the screen at y, so they cancel each other (as in Figure 4.5(4)). Because this cancelling occurs for $\Delta x = \lambda/2$ and we know that $\Delta s = d/2$, by substituting these into Equation 4.8 we find that the direction to the edge of the

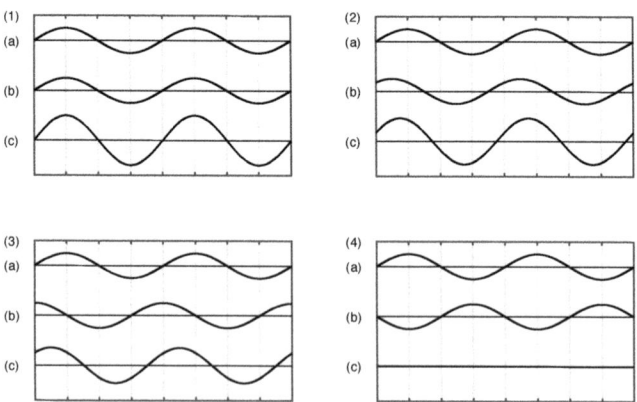

Figure 4.5: In each panel waves (a) and (b) are added to yield (c). The phase difference between (a) and (b) is: (1) $\Delta\theta = 0°$, so the waves are in phase and the amplitude of (c) is twice that of (a) and (b); (2) $\Delta\theta = 45°$; (3) $\Delta\theta = 90°$; (4) $\Delta\theta = 180°$, so the amplitude of (c) is zero.

diffraction envelope, where the light intensity is zero, is

$$\begin{aligned} \varphi &\approx (\lambda/2)/(d/2) \\ &= \lambda/d. \end{aligned} \tag{4.9}$$

Treating this angle φ as a measure of the amount of diffraction, it increases linearly with λ and decreases nonlinearly with the slit width d.

Even though every point s_T in the top half of the slit produces a wavelet in the direction φ, this wavelet will be cancelled by a wavelet from a point in the bottom half of the slit at $s_B = s_T - d/2$. The net result is that the intensity on the screen in the direction $\varphi = \lambda/d$ is zero. Therefore, since the edge of the central bright region is in the direction φ, its angular width is 2φ. From Equation 4.9, the image resolution can be increased by decreasing the wavelength λ or by increasing the aperture d.

For a circular aperture of diameter d, the Airy disc half-angle is $\varphi \approx 1.22\lambda/d$. If the imaging system has a lens then d refers to the diameter of the objective lens, which is why the best animal eyes have large pupils and the best telescopes have large objective lenses.

4.5. Diffraction and Uncertainty

Here, we first find an expression for the uncertainty Δy in position, then we find an expression for the uncertainty Δp_y in momentum and, finally, we confirm that their product $\Delta y \Delta p_y$ is at least $2h$, which is consistent with Heisenberg's uncertainty principle.

Position Uncertainty. As a photon passes through the slit in Figure 4.4, the uncertainty in position along the y-axis is the length of the slit,

$$\Delta y = d. \tag{4.10}$$

Therefore, position uncertainty can be reduced by decreasing d.

Momentum Uncertainty. Without loss of generality, we assume that photons have zero momentum along the y-axis as they arrive at the slit (this is equivalent to assuming that the light source is far away). But after passing through the slit, each photon travels toward the screen in a direction ϕ. Because almost all photons land within the central bright region between $\pm\varphi$, it follows that each photon's direction ϕ is almost certainly between $\pm\varphi$. For simplicity, we assume that the uncertainty $\Delta\phi$ in ϕ is the angular width of this bright region,

$$\Delta\phi \approx 2\varphi, \tag{4.11}$$

where (from Equations 4.9 and 4.10) $\varphi = \lambda/d$, so that

$$\Delta\phi \approx 2\lambda/d \tag{4.12}$$

$$= 2\lambda/\Delta y. \tag{4.13}$$

After passing through the slit, each photon can head off in a direction $\phi \neq 0$, so it can have a nonzero vertical momentum p_y. If the photon's total momentum is p then, using the small angle approximation,

$$\phi \approx p_y/p \text{ radians}, \tag{4.14}$$

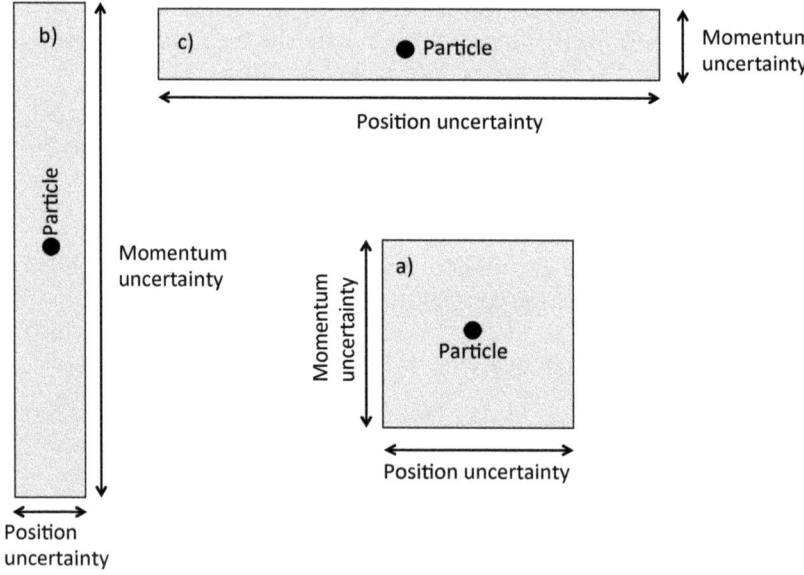

Figure 4.6: Geometric representation of uncertainty. For a particle, position uncertainty is represented on the horizontal axis and momentum uncertainty on the vertical axis, so the total uncertainty is the product of these uncertainties, which is represented by the grey area. a) Uncertainties in position and momentum have equal precision. b) Small position uncertainty means large momentum uncertainty. c) Small momentum uncertainty means large position uncertainty. The grey area, and hence the minimum uncertainty $2h$, is the same in a–c.

so that $p_y \approx p\phi$. This linear relationship tells us that uncertainty in ϕ is proportional to uncertainty in p_y, so the uncertainty in p_y is

$$\Delta p_y \approx p\Delta\phi. \tag{4.15}$$

Upon substituting $p = h/\lambda$ (Equation 4.1) and $\Delta\phi = 2\lambda/d$ (Equation 4.12), the uncertainty in momentum along the y-axis is

$$\begin{aligned} \Delta p_y &\approx (h/\lambda) \times (2\lambda/d) \\ &= 2h/d. \end{aligned} \tag{4.16}$$

Therefore, momentum uncertainty can be reduced by increasing d.

The product of the uncertainties in position (within the slit) and momentum (along the y-axis at the screen) is

$$\Delta y \, \Delta p_y \approx d \times 2h/d = 2h. \tag{4.17}$$

Thus, there is a trade-off between uncertainty in position and uncertainty in momentum, as shown in Figure 4.6. From Equation 4.10, we can decrease uncertainty in position by making the slit smaller, but this increases uncertainty in momentum (Equation 4.16). Conversely, from Equation 4.16, we can decrease uncertainty in momentum by making the slit larger, but this increases uncertainty in position (Equation 4.10). Crucially, we cannot reduce the joint uncertainty below $2h$.

A more intuitive picture is obtained by noting that (for small angles) $p_y \propto \phi$, so uncertainty in momentum implies uncertainty in the direction of a photon as it travels towards the screen (where direction uncertainty is proportional to the width of the central bright region in the diffraction pattern). And because each direction is uniquely associated with a position on the screen, the trade-off between position and momentum amounts to a trade-off between uncertainty in position within the slit and uncertainty in position on the screen.

The analysis above is a rough sketch, and Heisenberg's uncertainty principle was proved rigorously in 1927 by Kennard in the form

$$\sigma_y \sigma_p \geq \hbar/2 \ \text{J s}, \tag{4.18}$$

where σ_y and σ_p are the *standard deviations* of position and momentum, respectively (standard deviation is a statistical measure of uncertainty; see Glossary), and $\hbar = h/(2\pi)$ is the *reduced Planck's constant*.

For example, if the uncertainty in the position of an electron is $\sigma_y = 10^{-10}$ m (the diameter of a hydrogen atom) then the momentum uncertainty is at least $(\hbar/2)/10^{10}$, or 5.3×10^{25} kg m/s. Dividing this by the electron's mass (9.1×10^{31} kg), we obtain an enormous uncertainty in velocity, $\sigma_v = 5.8 \times 10^4$ m/s (about 0.2% of the speed of light).

Heisenberg's Inequality

This subsection is for readers familiar with *Fourier analysis*, which deals with the behaviour of waves. Because de Broglie's equation (Equation 4.1) allows each particle to be treated as a wave (more accurately, as a superposition of waves), the position and momentum of each particle can be defined in terms of Fourier analysis.

Briefly, Heisenberg's uncertainty principle is based on *Heisenberg's inequality*, which states that if a variable p_y with standard deviation σ_p is the *Fourier transform* of a variable y with standard deviation σ_y then the joint uncertainty $\sigma_y \sigma_p$ cannot be less than a critical value. For a particle with position y and momentum p_y, it can be shown [18] that Heisenberg's inequality implies Equation 4.18 (see Appendix F and Sanderson in Further Reading).

The Entropic Uncertainty Principle

Conventionally, uncertainty is expressed in terms of standard deviation, which is not a robust measure. In contrast, *Shannon's information entropy* [33] is a robust and universal measure of uncertainty. For brevity, we will refer to *Shannon's information entropy* as simply entropy. Somewhat confusingly, the terms entropy and information are used interchangeably, and both are measured in *bits*. Given two equally probable alternatives, the information required to choose the correct alternative is $\log_2 2 = 1$ bit (using logs to base 2 ensures that entropy is measured in bits). More generally, given N equally probable alternatives, the information required to choose the correct alternative is $H(N) = \log_2 N$ bits. This can be generalised further to define the information required to choose from an infinite number of equally probable alternatives in an interval on the real line, as in the following sketch of the entropic uncertainty principle.

Consider the position of a photon as it passes through a slit. In general, if a photon is equally likely to be at any position within a slit of width $d = \Delta y$ then the *differential entropy* in its position is

$$H(y) \quad = \quad \log_2 \Delta y \quad = \quad \log_2 d \text{ bits.} \tag{4.19}$$

Using entropy in place of standard deviation yields Beckner's (1975) *quantum entropic uncertainty principle* [3], which states that

$$\sigma_y \sigma_p \quad \geq \quad C\hbar/2 \quad \geq \quad \hbar/2, \tag{4.20}$$

where C is a measure of the entropy in y and p_y. If and only if the distributions of y and p_y are Gaussian then $C = 1$, which agrees with the standard inequality (Equation 4.18), but for all other distributions

$C > 1$. Crucially, these inequalities mean Beckner's entropic proof provides a tighter bound on uncertainty than Kennard's proof.

The width of the diffraction envelope in Figure 4.4 can be understood in terms of the entropic uncertainty principle. For position, the uncertainty is given by Equation 4.19. For momentum, the photon density on the screen is defined by a diffraction envelope of width 2φ radians. Provided the width of this envelope is small, the photon's momentum p_y is approximately proportional to its direction ϕ, and therefore $\Delta\phi \propto \Delta p_y$. The diffraction envelope is well approximated by a *Gaussian function* with width parameter (standard deviation) σ, as shown in Figure 4.7. Almost all the area under a Gaussian curve is within $\pm 2\sigma$ of its centre, so its effective width is about 4σ. By equating the widths of this Gaussian curve and the diffraction envelope, we obtain that $\Delta p_y \approx 4\sigma$ and therefore $\sigma \approx \Delta p_y/4$. This is useful because the entropy of a Gaussian[33] with standard deviation σ is $H = 0.5\log_2 2\pi e + \log_2 \sigma \approx 2.05 + \log_2 \sigma$, so that

$$H(p_y) \approx 2.05 + \log_2(\Delta p_y/4) \tag{4.21}$$
$$= 2.05 + \log_2 \Delta p_y - \log_2 4 \tag{4.22}$$

and hence, using $\log_2 4 = 2$,

$$H(p_y) \approx \log_2 \Delta p_y \tag{4.23}$$
$$= \log_2(2h/d) \text{ bits,} \tag{4.24}$$

where the final equality makes use of $\Delta p_y = 2h/d$ (Equation 4.16). Therefore, the photon's total Shannon entropy is

$$H(y) + H(p_y) \approx \log_2 d + \log_2(2h/d) \tag{4.25}$$
$$= \log_2(2h) \text{ bits,} \tag{4.26}$$

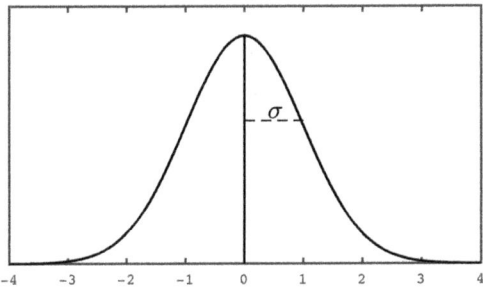

Figure 4.7: A Gaussian function with a standard deviation (width parameter) of $\sigma = 1$, as indicated by the horizontal dashed line.

which is consistent with Beckner's entropic uncertainty principle (this rough calculation only approximates Beckner's result). The fact that the entropy is less than zero is a trivial property of continuous variables[33].

If we now double the slit width, this doubles the uncertainty about the photon's position, which adds 1 bit to its position entropy. In order to respect the equality in Equation 4.26, adding 1 bit to the photon's position entropy must be compensated by removing 1 bit from its momentum entropy, which halves uncertainty about the photon's momentum. Provided the photon's angle ϕ is small, halving uncertainty in momentum translates directly into halving uncertainty $\Delta\phi$ about its angle ϕ. Consequently, we cannot reduce the total entropy below $\log_2(2h)$ bits. Finally, this is consistent with the fact that doubling the slit width roughly halves the width of the diffraction envelope, as implied by Equation 4.9.

Zero-Point Energy

The lowest energy level of a system, known as its *zero-point energy* or *ground state*, was initially the subject of much controversy. It was originally proposed by Planck in 1911, but by 1912 it was declared "dead as a doornail" by Einstein. However, in the same year, Einstein and Stern found that the average energy of a standing wave has a zero-point energy (*Nullpunktsenergie*) of $h\nu/2$.

Despite its name, the zero-point energy cannot be zero because if a particle is motionless then its momentum (and therefore its velocity) is zero, which means that the uncertainty Δp_y in its momentum is zero. Clearly, if either $\Delta y = 0$ or $\Delta p_y = 0$ then Heisenberg's uncertainty principle would be violated because in that case $\Delta y\,\Delta p_y = 0$.

4.6. Atomic Models

So far, we have seen how de Broglie's equation for matter waves underpinned Heisenberg's uncertainty principle. But it should always be remembered that these ideas were developed in the context of a physics community striving to make sense of new, and often counter-intuitive, experimental results. Of particular interest was the problem of the fine structure of matter. This programme of work has its roots in what is now known as *spectroscopy*, the study of which frequencies of light are absorbed by different elements. Spectroscopy played a key role in the first models of atomic structure. However, as we shall see, the final justification of these models depended on de Broglie's matter waves.

When light encounters hydrogen gas, most frequencies pass through unhindered, but a number of precisely defined frequencies are absorbed. Consequently, because the outermost region of the sun contains hydrogen, the spectrum of sunlight contains narrow dark *absorption lines*, called *Fraunhofer lines*. Similarly, when hydrogen is heated up, it emits light at particular frequencies, so the spectrum of hydrogen is dark except for a series of bright *emission lines*.

Each chemical element has its own unique set of emission lines, called a *spectral series*, which can be used to identify that element. For example, the first five emission lines of hydrogen occur at wavelengths of 656, 486, 434, 410 and 397 nm, as shown in Figure 4.8. The relationship between successive Fraunhofer lines is far from obvious, as can be seen from the wavelengths quoted above. However, by the late 1800s there was a strong suspicion that the spectral lines of an element were related to the internal structure of matter. Because light was known to consist of waves, it was thought that emission lines might be related to the harmonic series typical of wave-like behaviour.

As a first step to solving this problem, in 1885 a Swiss school teacher called Johann Balmer published an equation that produced values matching the wavelengths of the spectral series of hydrogen:

$$ \frac{1}{\lambda} \;=\; R \left(\frac{1}{4} - \frac{1}{n_2^2} \right) \; \mathrm{m}^{-1}, \tag{4.27} $$

where n_2 is the *principal quantum number* and R is the *Rydberg constant*. It was discovered later that the principal quantum number specifies the orbital 'height' of an electron (see Figure 4.9). The value of the Rydberg constant is $R = 1.097 \times 10^7 \ \mathrm{m}^{-1}$, which was estimated by fitting the wavelengths of emission lines to Balmer's formula. For $n_2 = 3\text{–}7$ this yields the first five wavelengths of the *Balmer series* shown in Figure 4.8. The reason for the subscript 2 in n_2 is explained in the next section.

Even though Balmer's formula was successful in predicting emission lines, it existed in a theoretical vacuum, free of any model for why emission lines should obey his formula. However, with the discovery of the electron by JJ Thomson in 1897, it became possible to theorise about how Balmer's formula might be related to the internal structure of atoms.

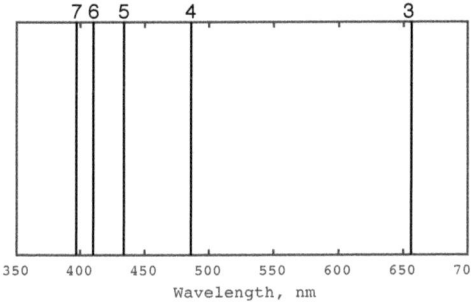

Figure 4.8: The Balmer series of lines for $n_2 = 3\text{–}7$ in Equation 4.27.

4.7. Bohr's Model

In 1911, Ernest Rutherford proposed the idea of a nuclear atom in which negatively charged electrons orbit a positively charged nucleus. But a fundamental problem was that an orbital trajectory involves continuous acceleration, which should cause the electron to lose energy and hence spiral down to the nucleus. Then, in 1913, Niels Bohr suggested a modification to Rutherford's model, which provided an important stepping stone along the path to quantum mechanics[5]. Crucially, Bohr proposed that an electron can revolve around the nucleus only in circular *stationary orbits*, where each orbit has an *orbital angular momentum* that is a multiple of $\hbar = h/(2\pi)$. The angular momentum of an electron with mass M_e moving at velocity v in a circular orbit with radius r is $M_e v r$. So Bohr's model required angular momentum to take only the discrete values

$$M_e v r \;=\; n\hbar \;\; \mathrm{kg\,m^2\,s^{-1}}, \qquad (4.28)$$

where $M_e = 9.109 \times 10^{-31}$ kg and $n = 1, 2, \ldots$ is related to the orbital 'height' of an electron. At a stroke, Bohr had solved (or avoided) the problem raised by the fact that orbiting electrons should lose energy. Inspired by previous work of Planck and Einstein, Bohr's proposal effectively assumed that electrons orbit the nucleus only at discrete radii, so the orbits have discrete energies. Bohr postulated that when an electron falls from an orbit with energy E_3 to a 'lower' orbit with energy E_2, a photon is emitted that has frequency

$$\nu \;=\; (E_3 - E_2)/h, \qquad (4.29)$$

in accordance with Planck's equation (Equation 2.21), as shown in Figure 4.9. Because orbits are assumed to occur only with discrete energies, movement of electrons between different orbits generates photons only with discrete frequencies, which correspond to emission lines.

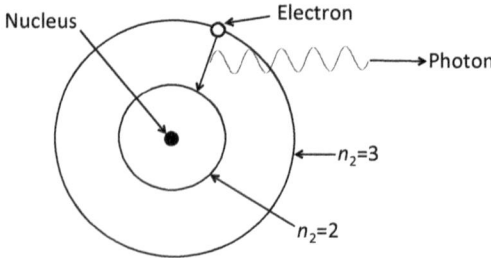

Figure 4.9: When an electron falls from the orbit with principal quantum number $n_2 = 3$ to the one with $n_2 = 2$, the energy ΔE released creates a photon with frequency $\nu = \Delta E/h$, where h is Planck's constant.

In the following sections, we first show that Bohr's assumption leads to a set of unique orbital radii, and we then show that the energy differences between these orbits generate photons at wavelengths that match the Balmer series. Because hydrogen is the simplest atom, with one negatively charged electron orbiting a positively charged proton (the nucleus), this was the atom Bohr used to test his theory.

Quantised Electron Orbits. To work out the unique radii of electron orbits, we need to know the relationship between the radius of an electron's orbit and the velocity of that electron. The force acting on a negatively charged electron in a circular orbit around a positively charged hydrogen nucleus is

$$F \;=\; M_e\, a = \frac{e^2}{4\pi\varepsilon_0 r^2}, \tag{4.30}$$

where M_e is the mass of an electron, a is the electron's inward acceleration, e is the charge on an electron, and ε_0 is the permittivity of space. The kinetic energy of the electron is

$$E_{\text{kinetic}} \;=\; M_e v^2/2, \tag{4.31}$$

and the acceleration is $a = v^2/r$. But from Equation 4.30,

$$a \;=\; F/M_e \;=\; \frac{e^2}{4\pi\varepsilon_0 r^2} \times \frac{1}{M_e}. \tag{4.32}$$

Putting these together, we have

$$\frac{v^2}{r} \;=\; \frac{e^2}{4\pi\varepsilon_0 r^2} \times \frac{1}{M_e}. \tag{4.33}$$

Solving for v^2 gives

$$v^2 \;=\; \frac{e^2}{4\pi\varepsilon_0 M_e} \frac{1}{r}. \tag{4.34}$$

It is worth noting that this is analogous to the relationship between the velocity v_{planet} of a planet in a circular orbit around the sun and the radius r_{planet} of the planet's orbit:

$$v_{\text{planet}}^2 \;=\; GM_{\text{sun}}/r_{\text{planet}}, \tag{4.35}$$

where M_{sun} is the sun's mass and G is the gravitational constant. For both electrons and planets, every value of the orbit's radius has a corresponding velocity, where velocity$^2 \propto 1/\text{radius}$.

At this point we have two constraints on the radius r, where both constraints can be expressed in terms of velocity. The first constraint is from the kinetic energy expressed in Equation 4.34, which implies

$$r = \frac{e^2}{4 M_e \pi \varepsilon_0} \times \frac{1}{v^2}, \tag{4.36}$$

plotted as the solid curve in Figure 4.10. As noted above, in the absence of any other constraints, every value of velocity has a corresponding radius, so we could always satisfy Equation 4.36 by trading radius against velocity.

The second constraint is from Bohr's assumption of discretised momentum, as expressed in Equation 4.28; this implies

$$r = \frac{n\hbar}{M_e} \times \frac{1}{v}, \tag{4.37}$$

which is plotted for $n = 1$ and $n = 2$ in Figure 4.10.

The two curves defined by Equations 4.36 and 4.37 intersect at a radius r that satisfies both constraints; for $n = 1$, this occurs at $r = 0.0529\,\text{nm}$. Thus, although each of the constraints in Equations

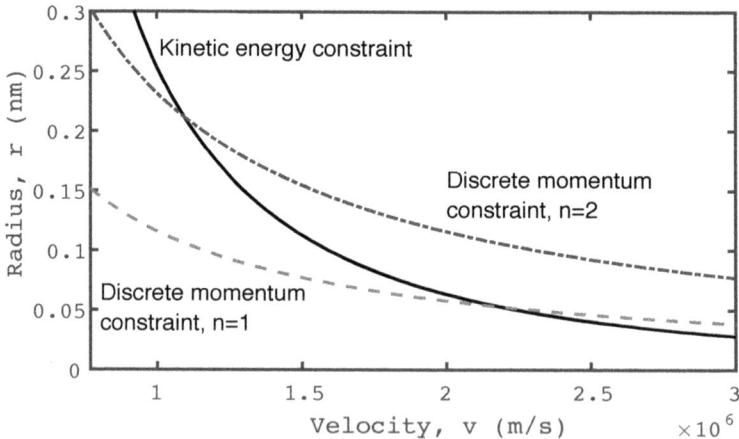

Figure 4.10: Constraints on the orbital radius of an electron. 1) Kinetic energy constraint: as velocity increases, the radius must decrease in accordance with Equation 4.34 (solid curve). 2) Discrete momentum constraint: Bohr's assumption that momentum exists in multiples of $nh/(2\pi)$ defines the dashed curve, for $n=1$ (Equation 4.37). Both constraints are satisfied for $n=1$ where the curves intersect at $r = a_0 = 0.0529\,\text{nm}$. Similarly, for $n = 2$ (dot-dashed curve), both constraints are satisfied at $r = 0.2117\,\text{nm}$. Velocity is in millions of m/s.

4.36 and 4.37 is satisfied by any value of r, both constraints are satisfied only at a single value of r. The value $r = 0.0529\,\text{nm}$ is known as the *Bohr radius*, which is taken to be the radius of a hydrogen atom and is traditionally denoted by a_0. Of course, the second constraint can be applied for any value of n, and Equation 4.37 for $n = 2$ is plotted as the dot-dashed curve in Figure 4.10. In this case, the intersection of the two constraint curves occurs at a radius of $r = 0.211\,\text{nm}$, which is the radius of the second stable orbit around the nucleus.

We can also calculate r algebraically. Equation 4.37 rearranges to

$$v = \frac{n\hbar}{M_e}\frac{1}{r}. \tag{4.38}$$

Squaring this and equating it to the right-hand side of Equation 4.34,

$$\frac{n^2\hbar^2}{M_e^2}\frac{1}{r^2} = \frac{e^2}{4\pi\varepsilon_0 M_e}\frac{1}{r}. \tag{4.39}$$

Solving for r yields

$$r = \frac{\hbar^2 4\pi\varepsilon_0}{M_e e^2}n^2. \tag{4.40}$$

For $n = 1$ this gives $r = 0.0529$ nm, which agrees with the value of the Bohr radius a_0 reported above. Notice that the Bohr radius is expressed entirely in terms of fundamental constants.

Bohr's Model and the Rydberg Constant. To work out the frequency of light emitted by an atom, Bohr needed to know the energy of each orbit in Equation 4.27. The total energy of an electron is the sum of its kinetic energy and potential energy, $E_{\text{tot}} = E_{\text{kinetic}} + E_{\text{potential}}$. The kinetic energy of an electron with mass M_e that is moving at v m/s is given in Equation 4.31. Substituting Equation 4.34 (v^2) into Equation 4.31 yields

$$E_{\text{kinetic}} = \frac{e^2}{8\pi\varepsilon_0}\frac{1}{r}. \tag{4.41}$$

The potential energy of an electron is

$$E_{\text{potential}} = \frac{-e^2}{4\pi\varepsilon_0 r}, \tag{4.42}$$

which is negative because, by convention, $E_{\text{potential}} = 0$ at $r = \infty$. In other words, $E_{\text{potential}}$ is the energy required to pull an electron out of its orbit to a distance of infinity. Therefore, the energy of an electron with an orbit of radius r is

$$E_{\text{tot}} = \frac{-e^2}{8\pi\varepsilon_0 r}. \tag{4.43}$$

Substituting Equation 4.40 (r) into Equation 4.43 yields

$$E_{\text{tot}} = \frac{-M_e e^4}{2(4\pi\varepsilon_0)^2 \hbar^2} \frac{1}{n^2} \tag{4.44}$$

$$= E_{\text{R}} \frac{1}{n^2} \text{ J}, \tag{4.45}$$

where

$$E_{\text{R}} = \frac{-M_e e^4}{2(4\pi\varepsilon_0)^2 \hbar^2} \text{ J} \tag{4.46}$$

is the *Rydberg energy constant*, whose value is 13.6 eV or 2.2×10^{-18} J. Therefore, the energy released when an electron falls from the orbit with *principal number* n_2 to the orbit with principal number n_1 is

$$\Delta E = E_2 - E_1 \tag{4.47}$$

$$= E_{\text{R}} \left(\frac{1}{n_1^2} - \frac{1}{n_2^2} \right). \tag{4.48}$$

As $\Delta E = h\nu = hc/\lambda$, the wavelength λ of the emitted radiation satisfies

$$\frac{1}{\lambda} = \frac{\Delta E}{hc} \tag{4.49}$$

$$= \frac{E_{\text{R}}}{hc} \left(\frac{1}{n_1^2} - \frac{1}{n_2^2} \right). \tag{4.50}$$

Inspired by Balmer's formula, Rydberg had first proposed this equation in 1888, expressed in terms of the *Rydberg constant R*

$$\frac{1}{\lambda} = R \left(\frac{1}{n_1^2} - \frac{1}{n_2^2} \right), \tag{4.51}$$

as a practical way to describe emission lines. In contrast, Bohr's analysis was firmly grounded in a theoretical framework, which was later extended by Wilson in 1915 and Sommerfield in 1916 to include elliptical orbits. The ratio E_{R}/hc equals the Rydberg constant,

$$R = \frac{-2M_e e^4}{(4\pi\varepsilon_0)^2 h^3 c} = 1.097 \times 10^7 \text{ m}^{-1}. \tag{4.52}$$

Before Bohr's model existed, the value of the Rydberg constant was estimated by fitting Balmer's equation (Equation 4.27) to emission line data. In contrast, Bohr's result expresses Rydberg's constant only in terms of universal physical constants. Moreover, Bohr's model provided a theoretical justification for Rydberg's formula and predicted the existence of three new line series, which were additions to the two known series of Balmer (for which $n_1 = 2$ in Equation 4.50) and Paschen (for which $n_1 = 3$). These three additional series were confirmed by Lyman ($n_1 = 1$), Brakett ($n_1 = 4$) and Pfund ($n_1 = 5$).

Bohr's model of the atom was a hybrid between the old world of classical physics and the nascent world of quantum mechanics. As such, some of its assumptions seemed justified, whereas others (e.g. quantised orbits) seemed entirely arbitrary. As shown in the next section, these apparently arbitrary assumptions would only start to make sense about a decade later in the context of de Broglie's matter waves.

4.8. Quantised Matter Waves

With the advent of de Broglie's proposal in 1924, it became possible to explain the form of Bohr's stationary orbits in terms of de Broglie wavelengths of electrons. In essence, Bohr's quantisation rules are equivalent to insisting that the orbit of each electron have a circumference that is equal to an integer number of de Broglie electron wavelengths (Figure 4.11). This is analogous to the problem in blackbody radiation of fitting an integer number of electromagnetic wavelengths into an oven (see Chapter 2), and is also an essential part of Schrödinger's wave mechanics introduced in the next chapter.

For an electron with mass M_e travelling at velocity v, its de Broglie wavelength is $\lambda_e = h/(M_e v)$. If only an integer number n of de Broglie wavelengths λ_e are permitted in a circular orbit with radius r then the

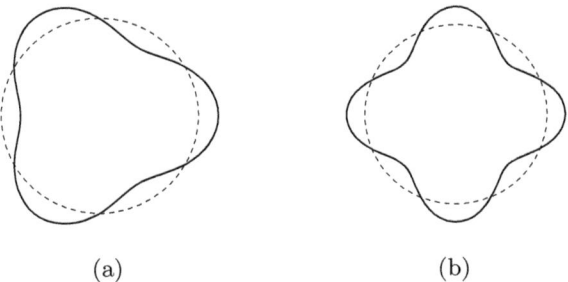

(a) (b)

Figure 4.11: Electron waves with integer numbers n of de Broglie wavelengths in imaginary two-dimensional atoms: (a) $n = 3$; (b) $n = 4$.

orbital circumference is

$$2\pi r = \lambda_e n \tag{4.53}$$

$$= hn/(M_e v), \tag{4.54}$$

and therefore

$$r = hn/(2\pi M_e v). \tag{4.55}$$

For $n = 1$ this gives the Bohr radius $r = a_0 = 0.0529$ nm (as reported on page 65). This implies an orbital circumference equal to a de Broglie wavelength of $\lambda_e = 2\pi a_0 \approx 0.33$ nm.

Squaring both sides of Equation 4.55, and substituting v^2 from Equation 4.34 gives

$$r^2 = \frac{h^2 n^2 4\pi\varepsilon_0 M_e r}{4\pi^2 M_e^2 e^2}. \tag{4.56}$$

Finally, simplifying yields Equation 4.40 as derived by Bohr, which can then be used to obtain the Rydberg equation (Equation 4.52). This represented a major success for quantum theory because, in contrast to the arbitrary assumptions of Bohr's model, the value of Rydberg's constant obtained here was based on the more realistic assumption that the circumference of an electron's orbit is restricted to values that must contain an integer number of de Broglie matter waves. Finally, we remark that the idea of an electron orbiting a nucleus like a planet around the Sun was a proxy for more sophisticated models of atomic structure based on de Broglie's matter waves.

4.9. Summary

In 1913, Bohr generalised Planck and Einstein's idea regarding the quantisation of light and applied it to electrons. Specifically, instead of adopting Einstein's assumption that light energy occurs in steps of size $h\nu$ joules, Bohr assumed that the angular momentum of electrons occurs in steps of size $h/(2\pi)$ kg m^2 s^{-1}. As a result, Bohr was able to derive Balmer's and Rydberg's formulae for spectral emission lines.

Almost a decade later, de Broglie combined Planck's equation relating frequency to energy with Einstein's equation relating mass to energy, to derive an equation expressing mass in terms of the de Broglie wavelength of matter waves. This led to the realisation that Bohr's ad hoc *assumption* regarding the quantisation of momentum was a *necessary condition* for the application of de Broglie's matter waves to the orbits of atomic electrons.

Chapter 5

The Double-Slit Experiment

*It contains the **only** mystery. We cannot make the mystery go away by 'explaining' how it works. We will just **tell** you how it works. In telling you how it works, we will have told you about the basic peculiarities of all quantum mechanics.*
Feynman R, 1965.

5.1. Introduction

By the year 1900, the debate regarding the nature of light had raged for over 300 years. The principal proponents of the *wave theory of light* were Christiaan Huygens (1629–1695) and Robert Hooke (1635–1703). In contrast, Isaac Newton (1642–1727) favoured the *corpuscular theory of light*, according to which light consists of tiny corpuscles, now called *photons*. Newton's pre-eminence ensured the dominance of his theory for several decades after his death.

Then, in 1801, Thomas Young described a simple experiment[37] at the Royal Society of London which seemed to prove conclusively that light consists of waves. The experiment performed by Young involved splitting a thin beam of sunlight along two sides of a thin piece of card; when the two half-beams recombined beyond the other end of the card, they produced a visible interference pattern on a screen. The *double-slit experiment* described below is the modern incarnation of Young's experiment. Just for reference, a double-slit experiment was

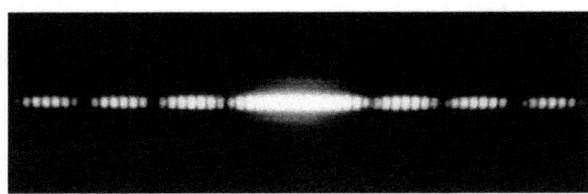

Figure 5.1: Typical interference pattern in the double-slit experiment shown in Figure 5.2.

used to produce the interference pattern shown in Figure 5.1, which is similar to the pattern observed in Young's experiment.

If we accept that light consists of waves then the fact that light from two slits interfere with each other seems unremarkable. But what if the light is so dim that it corresponds to a wave associated with a single photon? In 1909, Geoffrey Taylor (1886–1975) allowed the shadow cast by a fine needle to fall onto a photographic plate[35]. For bright light, its wave-like nature will allow it to diffract around the needle. The light from either side of the needle is analogous to light emanating from two slits, resulting in interference fringes in the needle's shadow, which were recorded on the photographic plate. In the context of Young's experiment, this was to be expected. What was not expected was what happened when the light was made to pass through glass blackened with carbon, so that only one photon hit the photographic plate at a time. Remarkably, after *2000 hours or about three months* of exposure time, the interference fringes on the photographic plate were found to be identical to those seen under normal lighting conditions. In other words, the waves associated with photons that reached the photographic plate one at a time produced the same interference pattern as obtained from the waves associated with streams of photons.

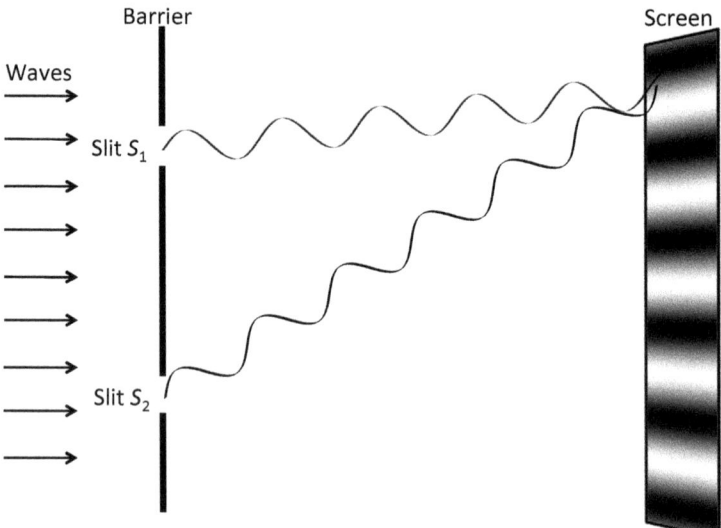

Figure 5.2: Waves travel from the left until they hit the barrier, which contains two slits, allowing a wave to emerge from each slit. These two waves interfere with each other and form an interference pattern on a screen (right). The waves shown are out of phase at a point on the screen, so they interfere destructively.

5.2. Interference in Light and Water

The double-slit experiment represents a distillation of all that is known about waves of light and matter. Fortunately, the mathematics that describes light waves has much in common with the mathematics of water waves. Accordingly, we begin with an account of the double-slit experiment in the context of water waves.

Consider a tank of water with a planar paddle at one end, which oscillates to produce regularly spaced waves on the surface of the water (see Figure 1.2). About half-way across the tank is a straight barrier that is parallel to the paddle and contains two vertical slits. When a wave hits the barrier, a small portion of the wave passes through each slit, and an approximately semi-circular wave emanates from each slit. Because these two waves originate from the same source, they are coherent (i.e. oscillate in synchrony) as they emerge from the slits. The waves then interfere with each other, producing distinct lines of peaks and troughs that radiate out from a single point mid-way between the slits. At the other end of the tank, which we will call the *screen*, the waves form an *interference pattern* (for simplicity, we assume that the screen absorbs energy, so that no waves rebound).

The water interference pattern is formed by simply adding the heights of the two waves at each screen position, a process known as *linear superposition*, as shown in Figure 4.5. This means that at any given point on the screen, if the two waves have the same phase then they reinforce each other, as in Figure 4.5(1). Conversely, if the two waves have different phases at that point then they cancel each other, either partially as in Figure 4.5(2) and (3), or completely as in Figure 4.5(4). When considered over the full extent of the screen, the result of the superposition of the two waves is the interference pattern shown in Figure 5.2. The intensity at a given point on the screen is the amount of energy delivered per second, and is proportional to the *squared height* of the wave at that point.

In the *light* double-slit experiment, a single light source is shone onto two narrow slits. The light emanating from the slits is *coherent*, which means that the waves are in phase with each other. As was the case for water, the intensity of the light interference pattern results from summing, and then squaring, the heights of the waves from the two slits at each screen position. Except for a change of spatial scale, the intensity of light and water at corresponding points in the two experiments are proportional to each other. Thus, aside from a few subtle differences (explored below), Figures 5.1 and 5.2 are an accurate representation of the interference patterns observed in both the water and the light double-slit experiments. This might sound a little strange, but not as strange as when we try to work backwards to find a physical correlate for what the light waves are made of.

Interference Bands and Diffraction Envelopes

The observant reader will have noticed that the light interference pattern in Figure 5.1 varies in brightness from left to right. This is because the interference pattern is the product of two components: *interference banding* and a *diffraction envelope*. The interference banding has been described briefly above and will be explored in more detail below. The origin of the diffraction envelope is explained here.

The diffraction envelope in the double-slit experiment is the sum of two diffraction patterns, one from each slit. Each diffraction pattern is identical to the single-slit diffraction pattern shown in Figure 4.4. If the two slits are sufficiently close then their corresponding diffraction patterns will form a single diffraction envelope, as shown in Figure 5.3a,b. In contrast, the interference banding is formed from the interference between waves from the two slits, as shown in Figures 5.2 and 5.3c. Thus, the observed interference pattern results from the combined effects of the interference banding and the summed diffraction patterns from the two slits (i.e. the diffraction envelope), as shown in Figure 5.3d.

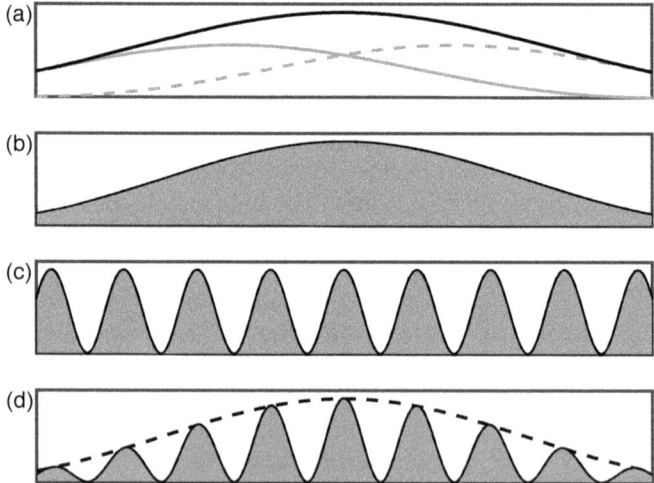

Figure 5.3: How the interference pattern from two slits is formed.
(a) The Fraunhofer diffraction envelopes from the two slits (grey dashed and grey solid curves, each the same as in Figure 4.4) combine to form a single diffraction envelope (black solid curve).
(b) The combined diffraction envelope, redrawn from (a).
(c) Interference between waves from both slits yields interference bands.
(d) The product (b) × (c) yields the interference pattern represented by the shaded region, which corresponds to Figure 5.1; the dashed curve is the combined diffraction envelope in (b).

5.3. Flying Clocks and Oscillators

In order to understand exactly how the double-slit experiment produces interference patterns, we first need to know about *travelling waves*. This entails some repetition of material presented in previous chapters, but it also provides a more general account.

Imagine a clock moving in a straight line along the x-axis, with its face in the direction of travel, as in Figure 5.4a. As the clock moves, the second hand traces out a spiral in space, where one complete rotation or *cycle* of the second hand defines one loop of the spiral.

For simplicity, we assume that the clock moves at constant speed but that the second hand can rotate at any rate. This means that the 'tightness' of the spiral depends only on how quickly the second hand rotates; the faster it rotates, the tighter the spiral. The constant speed of the clock corresponds to the constant speed of light, and the rate at which the second hand rotates corresponds to the angular frequency.

Every time the second hand rotates around the clock face through one cycle of 2π radians, it traces out one complete loop of the spiral. For example, if the second hand starts at 9 o'clock then this defines a point in 3-space (x, y, z), which we denote by (x_0, y_0, z_0). After one full rotation, the second hand returns to 9 o'clock (so y and z return to their original values). But in the time required for one cycle, the clock

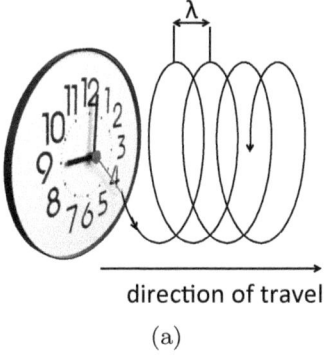
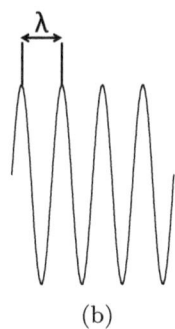

direction of travel

(a)   (b)

Figure 5.4: Geometric representation of a wave.
(a) A rotating second hand on a moving clock traces out a spiral in 3-space, with wavelength λ m. If the second hand completes ν rotations per second (one rotation is 2π radians) then it has an angular speed of $\omega = 2\pi\nu$ radians per second (rad/s). If the clock moves at c m/s then the number of wavelengths (spirals) that pass a fixed point each second is the frequency $\nu = c/\lambda$ Hz.
(b) Viewed from the side, the spiral defines a sinusoidal curve, which also has wavelength λ m and frequency ν Hz.

has moved along the x-axis to x_1, so now the second hand is at the point (x_1, y_0, z_0). The distance $x_1 - x_0$ travelled by the clock is equal to the wavelength λ of the spiral, which is the distance between successive loops of the spiral, as shown in Figure 5.4a.

If the second hand starts at 9 o'clock, and if the spiral casts a shadow on the ground, then one spiral loop corresponds to one cycle of a cosine wave on the ground, where the distance between consecutive peaks of the wave is the wavelength λ, as shown in Figure 5.5. Similarly, if the spiral casts a shadow on a vertical plane then the result is a sine wave with the same wavelength λ.

These different shadows of the spiral effectively provide two *orthogonal projections*, which correspond to the *real* and *imaginary* components of a *complex wave equation* (indicated by the hat symbol above Ψ):

$$\hat{\Psi} = A(\cos\theta + i\sin\theta) \tag{5.1}$$
$$= Ae^{i\theta}, \tag{5.2}$$

where θ is the angle between the second hand and the horizontal axis, A is the amplitude of the wave (which can be real or complex but is assumed to be real here), and $i^2 = -1$ (see Appendix D). Equation 5.2 defines a wave with squared amplitude

$$|\hat{\Psi}|^2 = \hat{\Psi}^* \times \hat{\Psi}, \tag{5.3}$$

where $\hat{\Psi}^* = Ae^{-i\theta}$ is the *complex conjugate* of $\hat{\Psi}$, so that

$$|\hat{\Psi}|^2 = A^2(e^{-i\theta} \times e^{i\theta}). \tag{5.4}$$

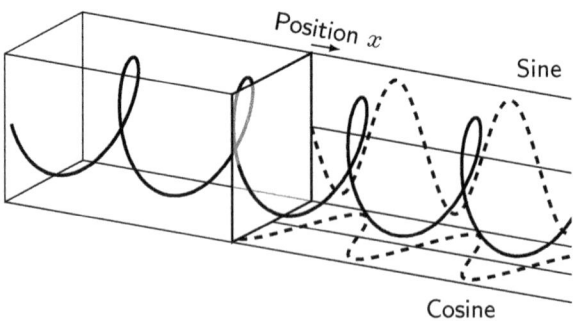

Figure 5.5: A complex wave $\hat{\Psi}$ represented as a spiral, which rotates around its central axis at the rate of ω radians per second (rad/s). In the box on the left, only the wave $\hat{\Psi}$ is shown. On the right, the real part of $\hat{\Psi}$ is the dashed cosine function on the ground plane, and the imaginary part is the dashed sine function on the vertical plane.

Since $e^{-i\theta} \times e^{i\theta} = 1$, the *amplitude* of $\hat{\Psi}$ is $|\hat{\Psi}| = A$.

If the wave travels at c m/s then the number of wavelengths that pass a fixed point in one second is its *frequency* $\nu = c/\lambda$ Hz (see Chapter 2). Because the wave travels λ metres for every complete cycle of 2π radians traced out by the second hand, its *wavenumber* is

$$k \quad = \quad 2\pi/\lambda \ \text{rad/m}. \tag{5.5}$$

At a distance of x metres from the origin, a sine wave with wavenumber k rad/m has a *spatial phase* of $\theta_x = kx$ radians.

A wave can also be described in terms of its *angular frequency*, which is the number of *radians per second* (rad/s) swept out by the second hand. The second hand sweeps out ν full rotations per second, so we can consider the clock to represent an *oscillator* with a frequency of ν Hz. The angular frequency is then $\omega = 2\pi\nu$ rad/s. If T seconds pass for every complete cycle of 2π radians then $T = 1/\nu = 2\pi/\omega$. We also know that in those T seconds the wave has travelled a distance of λ, so $T = \lambda/c$. Thus, the angular frequency is

$$\omega \quad = \quad 2\pi/T \quad = \quad 2\pi c/\lambda \ \text{rad/s}. \tag{5.6}$$

Note that the wavenumber (measured in rad/m) is the spatial analogue of the angular frequency ω (measured in rad/s). Thus, after t seconds, a wave with an angular frequency of ω rad/s has a *temporal phase* of $\theta_t = \omega t$ radians. In summary, the wave's phase changes with respect to location at the rate of k rad/m and with respect to time at the rate of ω rad/s.

5.4. The Travelling Wave

To explore the wave equation further, we will temporarily simplify notation by considering just the real part Ψ of the complex function $\hat{\Psi}$ in Equation 5.2. To avoid ambiguity, we refer to the value of Ψ for

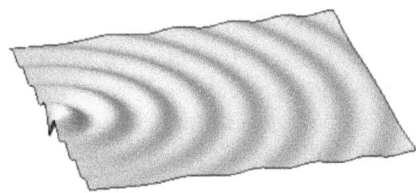

Figure 5.6: A circular travelling wave. Its phase θ at radius x and time t is a combination of its spatial phase θ_x at time $t_0 = 0$ and its temporal phase θ_t at $x_0 = 0$; specifically, $\theta = \theta_x - \theta_t$ radians.

a given value of θ as its *height*. Unlike with real waves, the height of every complex wave is its amplitude, which is constant over time.

The phase of a wave Ψ at a position x at time t consists of a spatial component θ_x and a temporal component θ_t. If we measure the distance x in wavelengths λ then the wave is x/λ wavelengths away from x_0. For now, assume that the wave at $x_0 = 0$ and time $t_0 = 0$ has a phase of zero radians, which means that the phase at position x and time t_0 is

$$\theta_x \;\; = \;\; 2\pi x/\lambda \;\; = \;\; kx \text{ radians.} \tag{5.7}$$

At any other time t, the phase at x will be shifted by some amount θ_t that depends on the angular frequency ω. Given that the angular frequency of the wave is $\omega = 2\pi/T$ rad/s, the temporal phase of the oscillator at time t is

$$\theta_t \;\; = \;\; 2\pi t/T \;\; = \;\; \omega t \text{ radians.} \tag{5.8}$$

If the wave is travelling to the right (i.e. with x increasing) at a speed of c m/s then at time t it has travelled $\Delta x = ct$ metres, so the phase at position x and time t equals the phase at position $x' = x - \Delta x$ and time t_0, which is

$$\begin{aligned} \theta \;\; &= \;\; (x - \Delta x)\,2\pi/\lambda \\ &= \;\; 2\pi x/\lambda - 2\pi ct/\lambda, \end{aligned} \tag{5.9}$$

where (from Equations 5.5 and 5.6) $k = 2\pi/\lambda$ and $\omega = 2\pi c/\lambda$, so that

$$\theta \;\; = \;\; kx - \omega t. \tag{5.10}$$

Thus, the height of a sinusoidal travelling wave varies as a function of position and time:

$$\begin{aligned} \Psi(x,t) \;\; &= \;\; A\cos(kx - \omega t) \tag{5.11} \\ &= \;\; A\cos(\theta_x - \theta_t). \tag{5.12} \end{aligned}$$

In summary, the function $\Psi(x,t)$ can be understood as a combination of two functions: a function that varies over position at each time, and a function that varies over time at each position. Specifically, at each position x, the function $\Psi(x,t)$ oscillates with amplitude A and a temporal period of $T = 2\pi/\omega$. Similarly, at each time t, the function $\Psi(x,t)$ oscillates with amplitude A and a wavelength of $\lambda = 2\pi/k$.

For completeness, the general sinusoidal travelling wave is often written as

$$\Psi(x,t) \;\; = \;\; C\cos(kx - \omega t) - D\sin(kx - \omega t), \tag{5.13}$$

where C and D are constants that determine the phase and the amplitude $A = (C^2 + D^2)^{1/2}$ of Ψ. For example, if $C = 0$ then $\Psi(x, t)$ is a sine function with an amplitude of $A = D$.

To return to the complex wave, substituting $\theta = kx - \omega t$ (from Equation 5.10) in Equation 5.2 yields a *wavefunction* with amplitude A,

$$\hat{\Psi}(x, t) = A e^{i(kx - \omega t)}, \tag{5.14}$$

so the complex wavefunction can be expressed as the product

$$\hat{\Psi}(x, t) = A e^{ikx} \times A e^{-i\omega t}, \tag{5.15}$$

which will prove useful later.

5.5. Real Interference

Here, we analyse a double-slit experiment that applies to water, as shown in Figure 5.7, and to electromagnetic waves, such as light. This section uses real numbers, so it applies to water waves; an analogous account using complex numbers, which applies to light waves, is presented in the next section.

As usual, there are two slits S_1 and S_2 from which waves emanate and travel to a screen, where the pattern produced by the waves can be observed. To work out how the interference pattern arises, we need to find the sum of two waves at every position on the screen. For convenience, we consider only position on the screen in one dimension, y, as indicated in Figure 5.7. In a typical double-slit experiment, the two waves have the same frequency and phase at the slits, which means that the light is *coherent*. In general, the slits are at different distances x_1 and x_2 from any given screen position y. However, if the distance from the slits to the screen is much larger than the distance between the slits then we can assume that $x_1 = x_2$ and define this distance to be x (this assumption is reasonable except when we wish to make use of the difference in path lengths $\Delta x = x_1 - x_2$). Similarly, we assume that the amplitudes (i.e. the maximum water wave heights) A_1 and A_2 of the waves from S_1 and S_2 are the same at y, and we define $A = A_1 = A_2$.

We will consider the interference pattern as a function of position y on the screen, so we need to express y in terms of the path length x_1 from S_1 to y and the path length x_2 from S_2 to y. For a path length difference $\Delta x = x_1 - x_2$ and a distance d between the slits, we see from Figure 5.7 that

$$\sin \varphi = \Delta x / d \tag{5.16}$$
$$= y/x, \tag{5.17}$$

so the path length difference at y is

$$\Delta x \;=\; yd/x. \tag{5.18}$$

Because φ is small, we can make use of the small-angle approximation $\sin \varphi \approx \varphi$ to obtain

$$\varphi \;\approx\; \Delta x/d \text{ radians}, \tag{5.19}$$

and if φ is small then $D \approx x$, so that $\varphi \approx y/D$ radians.

We now have enough information to work out the distance Δy between successive minima in the interference pattern. If $\Delta x = \lambda/2$ (i.e. if the difference in path lengths from S_1 and S_2 is exactly half of one wavelength) then (from Equation 5.19) the waves from the two slits in the direction $\varphi_{\min} = (\lambda/2)/d$ are always $180°$ out of phase. In other words, the direction $\varphi_{\min}$ points to the first minimum $y_{\min}$ above $y = 0$ in the interference pattern. By symmetry, $-\varphi_{\min}$ points to the first minimum

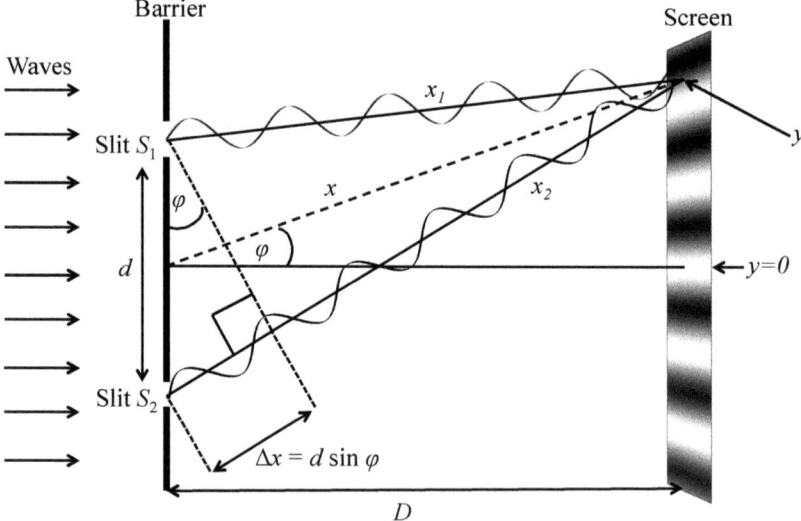

Figure 5.7: Geometry of a double-slit experiment. Waves with angular frequency ω and wavelength λ hit the barrier and escape as two waves Ψ_1 and Ψ_2 through the slits S_1 and S_2. The phase $\theta_1 - \omega t$ of Ψ_1 at S_1 and the phase $\theta_2 - \omega t$ of Ψ_2 at S_2 are identical. There is destructive interference at position y on the screen, because the path lengths x_1 and x_2 are different, which means that $\theta_1 = kx_1$ and $\theta_2 = kx_2$ are different. The phases of Ψ_1 and Ψ_2 at y change over time, but the *phase difference* $\Delta\theta = (\theta_1 - \omega t) - (\theta_2 - \omega t) = \theta_1 - \theta_2$ at y is constant. Note that if the distance D from the barrier to the screen is large then $x_1 \approx x_2$.

$-y_{\min}$ below $y = 0$; so the distance Δy between the two minima is $2y_{\min}$. Given that $\sin \varphi_{\min} = (\lambda/2)/d = y_{\min}/D$ and $\Delta y = 2y_{\min}$, the distance between minima is

$$\Delta y \;=\; \lambda D/d. \tag{5.20}$$

For example, if the distance between the slits is $d = 0.001\,\text{m}$, the distance from the slits to the screen is $D = 1\,\text{m}$ and the light wavelength is $\lambda = 500\,\text{nm}$ then the distance between consecutive minima (and between consecutive maxima) is $\Delta y = 0.5\,\text{mm}$.

If the angular frequency of the waves is ω rad/s then the phase at each slit at time t is ωt. The distance x_1 from slit S_1 to y measured in units of wavelengths λ is x_1/λ, which implies a phase of $\theta_1 = kx_1 = 2\pi x_1/\lambda$ radians at y. Notice that the value of θ_1 does not vary over time, so it is constant for a given screen position y. Therefore, if the phase is ωt at S_1 then it is $\theta_1 - \omega t$ at y. For example, if $\omega t = 0$ at S_1 then the phase is θ_1 at y. A similar argument applies to the phase θ_2 of the wave emanating from slit S_2.

Accordingly, for a given position y on the screen, the wave from slit S_1 has a height defined by the wavefunction

$$\begin{aligned}
\Psi_1(\theta_1, t) &= A\cos(kx_1 - \omega t) \\
&= A\cos(\theta_1 - \omega t),
\end{aligned} \tag{5.21}$$

where x_1, and therefore $\theta_1 = kx_1$, is a function of y. Similarly, the wave from slit S_2 is defined by the wavefunction

$$\Psi_2(\theta_2, t) \;=\; A\cos(\theta_2 - \omega t). \tag{5.22}$$

To find the interference pattern we need to add these waves at every position y on the screen. The sum of the wave heights at y is

$$\Psi(\theta_1, \theta_2, t) \;=\; A\cos(\theta_1 - \omega t) + A\cos(\theta_2 - \omega t). \tag{5.23}$$

Using the formula

$$\cos a + \cos b \;=\; 2\cos\frac{a - b}{2} \times \cos\frac{a + b}{2}, \tag{5.24}$$

Equation 5.23 becomes

$$\Psi(\theta_1, \theta_2, t) \;=\; 2A\cos\left(\frac{\theta_1 - \theta_2}{2}\right) \times \cos\left(\frac{\theta_1 + \theta_2}{2} - \omega t\right). \tag{5.25}$$

To tidy up notation, define $\Delta\theta = \theta_1 - \theta_2$ and $\bar{\theta} = (\theta_1 + \theta_2)/2$. Thus, at a given screen position y, $\Psi(\theta_1, \theta_2, t)$ is the product of a constant spatial component and a time-varying temporal component. The spatial component determines the amplitude of $\Psi(\theta_1, \theta_2, t)$ at y as $2A\cos(\Delta\theta/2)$. The temporal component modulates $\Psi(\theta_1, \theta_2, t)$ over time by a factor of $\cos(\overline{\theta} - \omega t)$, which varies between ± 1. Therefore, the value of $\Psi(\theta_1, \theta_2, t)$ at y varies over time between $\pm 2A\cos(\Delta\theta/2)$.

In general, intensity is energy per unit time and is obtained as the time-averaged wave height squared,

$$I(\Delta\theta, t) = 4A^2 \cos^2(\Delta\theta/2)\cos^2(\overline{\theta} - \omega t). \qquad (5.26)$$

The time-averaged value of $\cos^2(\overline{\theta} - \omega t)$ is 0.5, so intensity varies spatially as $I(\Delta\theta) = 2A^2 \cos^2(\Delta\theta/2)$ J/s. Note that the intensity of the real-valued wave in Equation 5.26 varies with the temporal phase. As we shall see, this is not true for waves associated with massive particles, such as electrons.

5.6. Complex Interference

This section provides a re-interpretation of the previous real analysis in terms of complex variables; it also prepares the way for the Schrödinger equation in Chapter 6.

Using $\theta_1 = kx_1$ and $\theta_2 = kx_2$ defined in the previous section, we assume that a complex wave $\hat{\Psi}_1(\theta_1, t)$ emanates from slit S_1 and another complex wave $\hat{\Psi}_2(\theta_2, t)$ emanates from slit S_2. From Equation 5.14,

$$\hat{\Psi}_1(\theta_1, t) = A_1 e^{i(\theta_1 - \omega t)}, \qquad (5.27)$$
$$\hat{\Psi}_2(\theta_2, t) = A_2 e^{i(\theta_2 - \omega t)}, \qquad (5.28)$$

where A_1 and A_2 are constants. The wavefunction at a given position on the screen is a linear superposition of these two waves,

$$\hat{\Psi}(\theta_1, \theta_2, t) = A_1 e^{i(\theta_1 - \omega t)} + A_2 e^{i(\theta_2 - \omega t)}, \qquad (5.29)$$

which can be written as a product of spatial and temporal components,

$$\hat{\Psi}(\theta_1, \theta_2, t) = (A_1 e^{i\theta_1} + A_2 e^{i\theta_2}) \times e^{-i\omega t} \qquad (5.30)$$
$$= \hat{\Psi}_s(\theta_1, \theta_2) \times e^{-i\omega t}, \qquad (5.31)$$

where $\hat{\Psi}_s(\theta_1, \theta_2)$ is a spatial wavefunction that depends only on position,

$$\hat{\Psi}_s(\theta_1, \theta_2) = A_1 e^{i\theta_1} + A_2 e^{i\theta_2}. \qquad (5.32)$$

Because both terms define waves that are sinusoidal, their sum must also be sinusoidal, so

$$\hat{\Psi}_s(\theta_1, \theta_2) = A_s e^{i\theta_s}, \qquad (5.33)$$

where A_s and θ_s are not known (yet). Thus, Equation 5.30 becomes

$$\hat{\Psi}(\theta_1, \theta_2, t) \quad = \quad A_s e^{i\theta_s} \times e^{-i\theta_t} \tag{5.34}$$

$$= \quad A_s e^{i(\theta_s - \theta_t)}. \tag{5.35}$$

In Equation 5.34, we have effectively isolated the spatial variation in $\hat{\Psi}$ as $\hat{\Psi}_s$ (because the spatial phase θ_s depends only on position) and the temporal variation as $e^{-i\omega t}$ (because the temporal phase θ_t depends only on time). Therefore, the phase of the complex wave $\hat{\Psi}$ at any given position is the sum of a spatial component with a constant phase θ_s and a temporal component with a phase θ_t that varies at ω rad/s.

At this stage, we could use these equations to work out how water waves behave, by taking the real part of Equation 5.35 and simply discarding the imaginary part of the complex wavefunction. This is usually done to avoid tedious algebra with sines and cosines, so it provides a convenient way to deal with wave equations. However, for the complex waves that describe light, we must retain both the real and the imaginary parts of the complex wavefunction.

We wish to find the intensity $I(y)\Delta y$ in a small region Δy centred on a screen position y. Before doing so, it will simplify matters to express this simply (but less accurately) as 'the intensity $I(y)\,\Delta y$ at y'. Notice two things. First, if $I(y)\Delta y$ is intensity then $I(y)$ should be an intensity density, but we will refer to it simply as intensity. Second, the screen position y determines two angles θ_1 and θ_2, as explained in the caption of Figure 5.7.

So, in order to find the intensity $I(y)$ at screen position y, we need to find the squared height of $\hat{\Psi}(\theta_1, \theta_2, t)$, which is

$$I(y) \quad = \quad |\hat{\Psi}(\theta_1, \theta_2, t)|^2 \tag{5.36}$$

$$= \quad \hat{\Psi}^*(\theta_1, \theta_2, t) \times \hat{\Psi}(\theta_1, \theta_2, t) \tag{5.37}$$

$$= \quad \hat{\Psi}_s^* e^{-i\omega t} \times \hat{\Psi}_s e^{i\omega t}. \tag{5.38}$$

Since $e^{-i\omega t} e^{i\omega t} = 1$,

$$I(y) \quad = \quad \hat{\Psi}_s^* \hat{\Psi}_s \tag{5.39}$$

$$= \quad |\hat{\Psi}_s|^2 \text{ J/s.} \tag{5.40}$$

Notice that, in the process of finding the intensity of the complex wavefunction $\hat{\Psi}(y)$, we lose all information regarding the temporal phase ωt of the wavefunction (as promised above). In other words, intensity varies with position but is constant over time.

From Equation 5.40, intensity is the squared amplitude of the complex wavefunction $\hat{\Psi}$ at y. Clearly, if the rate at which particles arrive at y is doubled then the intensity at y is also doubled. This, in turn, means that the probability $p(y)\,\Delta y$ that a particle will arrive at y in a given time interval is doubled; therefore, intensity is proportional to the probability density $p(y)$.

Expressing this as $p(y) \propto |\hat{\Psi}(y)|^2$ and using the fact that (by definition)

$$\int_y p(y)\,dy \;=\; 1, \tag{5.41}$$

we can ensure that the constant of proportionality is 1 by setting

$$\int_y |\hat{\Psi}(y)|^2\,dy \;=\; 1. \tag{5.42}$$

We can usually *normalise* the wavefunction to $\hat{\Psi}_{\text{normalised}}(y)$, such that

$$|\hat{\Psi}_{\text{normalised}}(y)|^2 \;=\; \frac{|\hat{\Psi}(y)|^2}{\int_y |\hat{\Psi}(y)|^2\,dy}, \tag{5.43}$$

which satisfies Equation 5.42. By convention, we assume that $\hat{\Psi}(y)$ is normalised (i.e. $\hat{\Psi} = \hat{\Psi}_{\text{normalised}}$) unless stated otherwise, so that

$$p(y) \;=\; |\hat{\Psi}(y)|^2 \tag{5.44}$$

and therefore

$$|\hat{\Psi}(y)| \;=\; \sqrt{p(y)}. \tag{5.45}$$

In words, the amplitude of the (normalised) wavefunction at the screen position y equals the square root of the probability density $p(y)$ at y.

We can evaluate the intensity of $\hat{\Psi}$ at y from the squared amplitude of $\hat{\Psi}_s$ (Equation 5.40). The intensity $|\hat{\Psi}_s|^2$ is obtained by multiplying $\hat{\Psi}_s$ by its complex conjugate (from Equation 5.32):

$$\begin{aligned}
I(y) \;&=\; \hat{\Psi}_s^* \hat{\Psi}_s & (5.46)\\
&=\; (A_1 e^{i\theta_1} + A_2 e^{i\theta_2}) \times (A_1 e^{-i\theta_1} + A_2 e^{-i\theta_2}) & (5.47)\\
&=\; [A_1^2 e^{i\theta_1} e^{-i\theta_1} + A_2^2 e^{i\theta_2} e^{-i\theta_2}]\\
&\quad + [A_1 e^{i\theta_1} A_2 e^{-i\theta_2} + A_2 e^{i\theta_2} A_1 e^{-i\theta_1}]; & (5.48)
\end{aligned}$$

the exponents in the first set of square brackets cancel, and the second set of brackets contain the cross terms, so

$$I(y) \;=\; A_1^2 + A_2^2 + A_1 A_2 [e^{i(\theta_1 - \theta_2)} + e^{i(\theta_2 - \theta_1)}]. \tag{5.49}$$

As before, we assume $A_1 = A_2$ and then define $A = A_1 = A_2$, so that

$$I(y) \quad = \quad 2A^2 + A^2[e^{i(\theta_1 - \theta_2)} + e^{i(\theta_2 - \theta_1)}]. \tag{5.50}$$

Defining the phase difference $\Delta\theta = \theta_1 - \theta_2$, we have

$$I(y) \quad = \quad 2A^2 + A^2[e^{i\Delta\theta} + e^{-i\Delta\theta}]. \tag{5.51}$$

Now we make use of the fact that

$$
\begin{aligned}
e^{i\Delta\theta} + e^{-i\Delta\theta} \quad &= \quad (\cos\Delta\theta + i\sin\Delta\theta) + (\cos\Delta\theta - i\sin\Delta\theta) \\
&= \quad 2\cos\Delta\theta \tag{5.52}
\end{aligned}
$$

to rewrite Equation 5.51 as

$$
\begin{aligned}
I(y) \quad &= \quad 2A^2 + 2A^2 \cos\Delta\theta \tag{5.53} \\
&= \quad 2A^2[1 + \cos\Delta\theta]. \tag{5.54}
\end{aligned}
$$

Finally, using the identity $\cos\Delta\theta = 2\cos^2(\Delta\theta/2) - 1$, the intensity is

$$
\begin{aligned}
I(y) \quad &= \quad |\hat{\Psi}(\theta_1, \theta_2, t)|^2 \tag{5.55} \\
&= \quad 4A^2 \cos^2(\Delta\theta/2), \tag{5.56}
\end{aligned}
$$

which is bounded between 0 and $4A^2$. Note that, in contrast to the real wavefunction Ψ (Equation 5.26), the intensity of this complex wavefunction does not vary with time.

We can explore what happens when only one slit is left open by using Equation 5.52 to rewrite Equation 5.49 as

$$I(y) \quad = \quad A_1^2 + A_2^2 + 2A_1 A_2 \cos\Delta\theta. \tag{5.57}$$

With both slits open, the intensity at any given location y on the screen consists of three components (Equation 5.57):

(1) the intensity $I_1 = A_1^2$ obtained with only slit S_1 open;

(2) the intensity $I_2 = A_2^2$ obtained with only slit S_2 open;

(3) an interference term $2A_1 A_2 \cos\Delta\theta$, which depends on the phase difference $\Delta\theta$ between the waves from S_1 and S_2 and varies between $\pm 2A_1 A_2$.

Bounds for the overall intensity can be obtained by rewriting $I(y)$ analogously to Equation 5.56.

If only slit S_2 is closed then $A_2 = 0$, so the intensity is $I_1 = A_1^2$. Similarly, if only slit S_1 is closed then $A_1 = 0$, so the intensity is $I_2 = A_2^2$. More importantly, if only one slit is open then the pattern is no longer an interference pattern.

5.7. What Is Seen in Light and Water

In the water double-slit experiment, the interference pattern we see is just the *time-varying height* of the wave at each point on the screen. In the light double-slit experiment, we do not see any quantity that corresponds to water height, but we do see light intensity, as shown in Figure 5.1. Because light intensity at any screen position y is proportional to the number of photons, it is also proportional to the probability that a photon will land at y. Consequently, the interference pattern of light intensity is identical to the density of photons shown in Figure 5.1, which is identical to the pattern of photon probabilities in Figure 1.3.

Intensity is proportional to the square of some other quantity, so (obviously) this quantity must be proportional to the square root of intensity. For water, the square root of intensity at y is related to the wave amplitude at y. But for light (and this is a crucial point), intensity is proportional to the probability that a photon lands at y, so *the square root of intensity at y is the square root of the probability that a photon will land at y* (Equation 5.45). This strange quantity ($|\hat{\Psi}(\theta_1, \theta_2)|$), the square root of a probability, is called the *probability amplitude*.

5.8. Which Slit for Which Photon?

In the double-slit experiment with water, there is no question of which slit a particle passes through, because there is no particle to be considered. The water is simply a medium, consisting of many molecules, that transmits a wave through both slits. In the light double-slit experiment, a wave of light passes through both slits, and the resultant waves produce an interference pattern on the screen. Incredibly, even when the light source is made so dim that only one photon reaches the screen at a time, the interference pattern still appears (albeit very slowly, as shown in Figure 4.5). In this case, it is as if a wave that corresponds to a single particle passes through both slits and then interferes with itself to produce an interference pattern on the screen. But a photon, unlike a wave, cannot pass through both slits, which raises the question: did the photon really pass through only one slit, and if so, which one?

To attempt to answer this question, consider what happens if we replace the screen with an array of long tubes, each of which points at just one slit, as shown in Figure 5.8a. At the end of each tube is a photodetector, such that any photon it detects could have come from only one slit. Note that there should be a pair of detectors at every screen position, with each member of the pair pointing at a different slit. Thus, irrespective of where a photon lands on the screen, the slit from which it originated is measured.

If we were to use this imaginary apparatus then we would find that the distribution of photons is a diffraction envelope, as in Figures 5.3a, 5.8a and 5.9a. Crucially, this envelope is identical to the pattern that would be obtained if the slits were opened one at a time. Specifically, first one slit is opened and the image on the screen is recorded, then the other slit is opened and the image is recorded, and finally these images are added together. In other words, using detectors to measure each photon's *slit identity* (i.e. which slit the photon passed through) prevents any wave-like behaviour, just as if each photon had travelled in complete isolation as a single particle. If both slits are left open (and no photodetectors are used) then the original interference pattern is restored, as if the individual photons behave like waves (as in Figures 5.3d and 5.8b). This is the famous *wave–particle duality*.

As counter-intuitive as this result seems, it is consistent with Heisenberg's uncertainty principle (see Sections 4.3 and 4.4). This ensures that when the slit identity is not measured, uncertainty in the particle's screen position is roughly equal to the distance between successive maxima in the interference pattern[6;18] (i.e. it is of the same order of magnitude as the width of each bright region in Figure 5.7). Note that the position of a particle on the screen is determined by that particle's momentum as it exits a slit, so screen position translates to momentum at the barrier containing the slits; similarly, slit identity translates to the position of the particle at the barrier (see Section 4.4). Therefore, any reduction in the uncertainty in slit identity (position) must increase uncertainty in the screen position (momentum)[26;27]. It is possible to adjust the amount of information gained regarding position by varying the accuracy of the measurement devices. As more information on position is gained, less information on the fine structure of the interference pattern is available. Consequently, as more information about position is gained, the interference pattern (Figure 5.9b) is gradually replaced by the sum of two broad diffraction envelopes[19] (Figures 5.3b and 5.9a). (For an extended discussion, see Feynman[12], Vol. I, Ch. 3.)

5.9. Wheeler's Delayed-Choice Experiment

So far, we have chosen to measure two different aspects of each photon: first, measuring each photon's position on the screen (which translates into momentum at a slit; see previous section); and second, using tube detectors to measure which slit each photon came from (which translates into the photon's position at the barrier). Clearly, when measuring photon position, the experimental setup does not change over time, so it seems plausible (or at least acceptable) that each photon could have passed through both slits. Similarly, when measuring slit identity, it seems plausible that each photon passes through only one slit. But there is an alternative experiment, which involves changing the experimental setup while each photon is in transit between the slits and the screen. If what we choose to measure alters how each photon behaves then it seems reasonable that we must make this decision before each photon reaches the slits. However, what if the decision on whether to measure screen position or slit identity is made *after* each photon has passed through the slit(s) but before it has reached the screen or tube detectors? This is *Wheeler's delayed-choice experiment*, depicted in Figure 5.8.

Such an experiment is conceptually easy to set up. Once the photon is in transit between the slits and the screen, we can decide whether to measure the photon's screen position (by leaving the screen in place, Figure 5.8b) or slit identity (by removing the screen so that the tube detectors can function, Figure 5.8a).

The results of an experiment essentially no different from this were published in 2007[17]. When the screen was left in place, the photons'

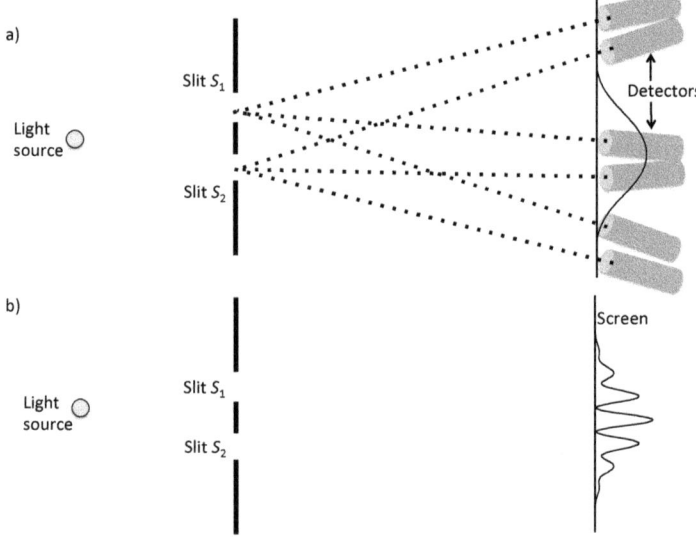

Figure 5.8: An imaginary experiment for demonstrating wave–particle duality. a) If slit identity (photon position at the barrier) is measured using an array of oriented detectors at the screen then momentum (direction) precision is reduced, and a diffraction envelope is observed on the screen, as here and in Figure 5.9a. b) If slit identity is not measured then the screen is allowed to measure direction (photon momentum) with high precision, so an interference pattern is observed, as here and in Figure 5.9b. In Wheeler's delayed-choice experiment, we decide to measure either a) slit identity, or b) photon momentum, but the decision is made after the photon has passed through the slit(s).

screen positions were effectively measured, which yielded an interference pattern on the screen. In contrast, when the screen was removed, an array of detectors was revealed that detected photons consistent with two diffraction envelopes (one per slit), as if no interference had occurred. Crucially, in both cases, the decision on measuring screen position or slit identity was made (at random) after each photon had passed through the slit(s), so the behaviour of each photon as it passed through the slit(s) depended on a decision made after that photon had passed through the slit(s). In essence, it is as if a decision made now about whether to measure the slit identity or screen position of a photon (that is already in transit from the slits to the screen) retrospectively affects whether that photon passed through just one slit or both slits.

In principle, the slit–screen distance can be made so large that it takes billions of years for each photon to travel from the slit(s) to the screen. In this case, a decision made now about whether to measure

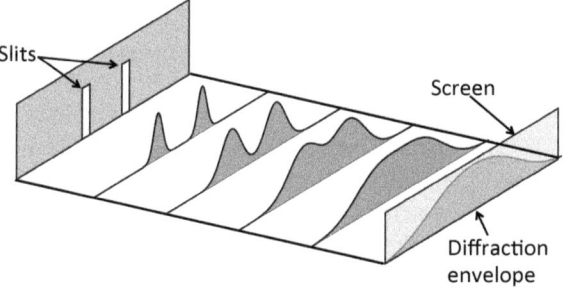

(a) Measuring position (slit identity) at the slits

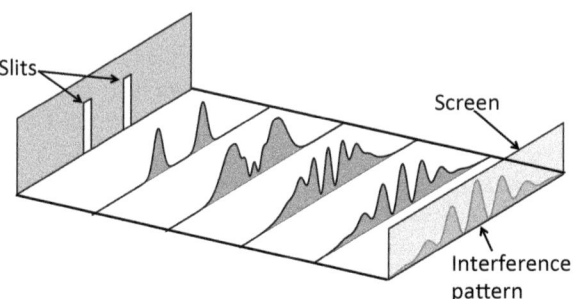

(b) Measuring momentum (direction or screen position)

Figure 5.9: How probability density evolves over distance after photons pass through two slits [19]. (a) If slit identity (photon position) is measured then the waves from the two slits do not interfere with each other. (b) If slit identity is not measured then the waves from the two slits produce an interference pattern as they travel further from the slits. For display purposes, the distribution at each distance has unit height.

the slit identity or screen position of a photon seems to retrospectively affect whether that photon passed through just one slit or both slits at once *billions of years ago.*

As we should expect, these results are consistent with Heisenberg's uncertainty principle. Regardless of when the decision is made, if the detectors measure slit identity (position) then this must increase uncertainty regarding the particle's screen position (momentum), which leads to disappearance of the interference pattern. Even though it is far from obvious how any physical mechanism could produce this result, the fact remains that if such a mechanism did not exist then Heisenberg's uncertainty principle would be violated.

5.10. Summary

The double-slit experiment represents a brutally harsh test for examining our intuitions about quantum mechanics. If we think we have an idea about how light, electrons or molecules should behave, either as particles or as waves, then the double-slit experiment will test that idea rigorously — and will invariably test it to destruction.

So, does the double-slit experiment allow us to conclude that a given entity is a particle or a wave? No. In accordance with Bohr's definition of *complementarity*, it only allows us to conclude that an electron sometimes behaves like a particle (e.g. when it crashes into a detector) and sometimes behaves like a wave (e.g. when it seems to pass through two slits at the same time). Alternatively, we could adopt de Broglie's hypothesis that an electron is guided by a *pilot wave*. And the most we can say about this pilot wave is that it seems to be made of something like the square root of probability.

In conclusion, we have equations that accurately describe the behaviour of photons and electrons. However, whereas Newton's equations describe the trajectory of a planet, so that we can know the exact position of Earth in 1000 years' time, the corresponding equations of quantum mechanics describe only the exact *probability* of every possible trajectory of a particle. Consequently, we can only know the probability that an electron will be in a particular position at a particular time. And even though these probabilities are known exactly, they are, nevertheless, just probabilities.

The gap between the precise predictions of position within classical physics and the precise predictions of probability for each possible position provided by quantum mechanics worried many physicists, especially Einstein, who famously claimed that "God does not play dice". To this day, it is widely accepted that such a gap means that quantum mechanics, for all of its many triumphs, is incomplete.

Chapter 6

Schrödinger's Wave Equation

Our imagination is stretched to the utmost, not, as in fiction, to imagine things which are not really there, but just to comprehend those things which are there.
Feynman R, 1967.

6.1. Introduction

Schrödinger's wave equation defines a wave that varies over space and time[30]. However, whereas a wave of water is defined by the varying height of water, the wave defined by Schrödinger's equation is something like a wave of information. In essence, Schrödinger's equation carries all of the information necessary to specify the probability that physical measurements (e.g. of position) will yield particular values at any given time and place.

We should make it clear from the outset that Schrödinger did not derive his equations in the usual manner. Instead, he considered what type of equation could explain the observed behaviour of particles in order to come up with an equation that might fit such behaviour. In this respect, Schrödinger's equation is not the result of conventional scientific deduction. Just as Planck's equation for blackbody radiation was an act of desperation, so Schrödinger's equation was an act of imaginative creation.

6.2. Quantum Guitar Strings

This section is a rough sketch, a conceptual scaffold, intended to act as a supporting framework for Schrödinger's quantum mechanical edifice.

As described in Chapter 2, when a guitar string is plucked, a whole series of frequencies is generated, as shown in Figure 2.3. But this series includes only particular frequencies, starting with the fundamental frequency, plus an infinite number of harmonics. Crucially, the fundamental frequency depends on the length of the guitar string and has a wavelength that is twice the length of the guitar string. Because the string is fixed at both ends, the fundamental frequency and the frequencies of the harmonics are each associated with a standing wave, such that each standing wave has an integer number of half-wavelengths. Thus, a guitar string is constrained to generate a particular *discretised* set of frequencies.

So far, the guitar string is an excellent model for an electromagnetic wave that is confined to a cubic oven. This model becomes more exact if we restrict attention to a line along one side of the oven, so that the oven is essentially one-dimensional. Just as each frequency of a standing wave on a guitar string corresponds to an integer number of half-wavelengths, so each frequency of an electromagnetic standing wave in a one-dimensional oven corresponds to an integer number of half-wavelengths. Thus, whether a standing wave is on a guitar string or in a one-dimensional oven, it can only adopt a particular discretised set of frequencies. However, if we assume that the guitar exists in the world of classical physics then a standing wave on a guitar string can adopt any amplitude (loudness), whereas an electromagnetic standing wave cannot. This is where the (classical) guitar string standing wave and the (quantum mechanical) electromagnetic standing wave part company.

Unlike waves on a guitar string, all electromagnetic standing waves in an oven with the same frequency have exactly the same energy, which means that they all have exactly the same amplitude. More precisely, the energy of electromagnetic standing waves can *only* adopt values that are multiples of a *quantum* of energy, where one quantum is the product of the standing wave's frequency and Planck's constant. In modern terminology, this product is the energy of a single photon.

In summary, the standing waves on a guitar string are quantised because they can adopt only certain frequencies, but they can adopt any amplitude at each of these frequencies, with each amplitude defining a different energy level. In contrast, electromagnetic standing waves are *doubly* quantised: not only are they constrained to adopt certain frequencies, but each frequency can only adopt discrete energy levels, with each energy level corresponding to an integer number of photons. Thus, a guitar string is quantised in frequency but not energy, whereas a quantum system is quantised in both frequency and energy. It is this energy quantisation that is unique to quantum mechanics.

In the following sections, we define the classical wave equation as a differential equation and then use a particular type of solution to that equation (i.e. standing waves) as a template to obtain the Schrödinger wave equation.

6.3. The Classical Wave Equation

Almost everything in quantum mechanics is expressed in terms of waves. Because the most fundamental wave is the sinusoid, we begin with an overview of sinusoidal waves. Consider a rope held at one end, which is being raised and lowered about once a second, generating a wave that travels along the rope. The vertical displacement $\Psi(x, t)$ of this wave at position x and time t is described by a second-order partial differential equation, which is known as the *classical wave equation*:

$$\frac{\partial^2 \Psi(x, t)}{\partial x^2} = \frac{1}{v^2} \frac{\partial^2 \Psi(x, t)}{\partial t^2}, \tag{6.1}$$

where v is the *phase velocity* (the speed at which the wave travels along the rope). This equation has an infinite number of solutions, but we focus on sinusoidal solutions. Incidentally, the classical wave equation would normally be introduced over several weeks in an undergraduate physics course, but it is presented without proof here.

The rope goes all the way up and down in T seconds, which corresponds to an oscillator that sweeps out 2π radians every T seconds, so the angular frequency is

$$\omega = 2\pi/T \text{ rad/s.} \tag{6.2}$$

Similarly, if the wave travels λ metres every time the oscillator cycles through 2π radians then the wavenumber is $k = 2\pi/\lambda$ rad/m. Note that the phase velocity is often expressed as

$$v = \omega/k \text{ m/s,} \tag{6.3}$$

As stated in Section 5.4, the height of the travelling wave at position x and time t is defined by the wavefunction

$$\Psi(x, t) = \cos(kx - \omega t). \tag{6.4}$$

The fact that this is a solution to the classical wave equation (Equation 6.1) can be checked by differentiating it as specified in that equation, as will be shown in Section 6.4. It will prove useful later to rewrite Equation 6.1 in terms of complex numbers, as a complex classical wave equation

$$\frac{\partial^2 \hat{\Psi}(x, t)}{\partial x^2} = \frac{1}{v^2} \frac{\partial^2 \hat{\Psi}(x, t)}{\partial t^2}, \tag{6.5}$$

where the hat symbol denotes a complex function. By analogy with Equation 6.4, solutions to the complex wave equation are of the form

(see Appendix D and Section 5.4 with $A = 1$)

$$\hat{\Psi}(x,t) \quad = \quad e^{i(kx - \omega t)}. \tag{6.6}$$

As with Equation 6.4, the fact that this is a solution to the complex classical wave equation (Equation 6.5) can be checked by differentiating it as specified in the complex classical wave equation, as will be shown in Section 6.5.

6.4. Stationary Classical Waves

Now, rather than considering a travelling wave, we constrain the wave so that it is fixed at both ends, like a guitar string. In this case, solutions to the classical wave equation (Equation 6.1) correspond to *standing waves* or *normal modes*, as shown in Figure 2.3.

For ease of reference, we assume that the guitar string is horizontal and that oscillations are variations in height over time. Within each standing wave, all points on the guitar string move in synchrony but with an amplitude that varies with position. For standing waves, the function $\Psi(x,t)$ can be separated into the product of two functions, $f(x)$ and $g(t)$, where $f(x)$ is a function of position only and $g(t)$ is a function of time only:

$$\Psi(x,t) \quad = \quad f(x)\, g(t). \tag{6.7}$$

The function $f(x)$ defines how the amplitude varies with position x, and $g(t)$ defines how the vertical height at each position changes over time. Using Equation 6.7 to express a solution to Equation 6.1 as a product of separate functions of position and time means that we can evaluate the left- and (then) right-hand sides as follows.

First, for the left-hand side of Equation 6.1, we find the second derivative of $\Psi(x,t)$ with respect to x. The first derivative is

$$\frac{\partial \Psi(x,t)}{\partial x} \quad = \quad g(t)\frac{df(x)}{dx} + f(x)\frac{dg(t)}{dx}. \tag{6.8}$$

Since $dg(t)/dx = 0$, the final term in Equation 6.8 equals zero, so the second derivative of $\Psi(x,t)$ is

$$\frac{\partial^2 \Psi(x,t)}{\partial x^2} \quad = \quad g(t)\frac{d^2 f(x)}{dx^2} + \frac{dg(t)}{dx}\frac{df(x)}{dx}. \tag{6.9}$$

Again, the final term equals zero, which leaves

$$\frac{\partial^2 \Psi(x,t)}{\partial x^2} \quad = \quad g(t)\frac{d^2 f(x)}{dx^2}. \tag{6.10}$$

For the right-hand side of Equation 6.1, we find the second derivative of $\Psi(x,t)$ with respect to t. By symmetry,

$$\frac{\partial^2 \Psi(x,t)}{\partial t^2} = f(x)\frac{d^2 g(t)}{dt^2}. \tag{6.11}$$

Substituting Equations 6.10 and 6.11 into Equation 6.1 and then rearranging yields a classical wave equation for standing waves,

$$\frac{1}{g(t)}\frac{d^2 g(t)}{dt^2} = \frac{v^2}{f(x)}\frac{d^2 f(x)}{dx^2}. \tag{6.12}$$

We now have two functions of two different variables x and t,

$$F(x) = \frac{v^2}{f(x)}\frac{d^2 f(x)}{dx^2} \tag{6.13}$$

and

$$G(t) = \frac{1}{g(t)}\frac{d^2 g(t)}{dt^2}, \tag{6.14}$$

such that $F(x) = G(t)$. The only way that two functions of different variables x and t can be equal for all values of x and t is if $F(x)$ and $G(t)$ are equal to the same constant, called the *separation constant*. Here, we call that constant $-\omega^2$, where (from Equation 6.3) $\omega = vk$ with k being the wavenumber. Next, we find expressions for $f(x)$ and $g(t)$.

Evaluating $f(x)$. Substituting $\omega = vk$ and $F(x) = -\omega^2$ into Equation 6.13 and rearranging yields

$$\frac{d^2 f(x)}{dx^2} = \frac{-\omega^2}{v^2}f(x) \tag{6.15}$$

$$= -k^2 f(x), \tag{6.16}$$

which has sinusoidal solutions of the form

$$f(x) = C\cos(kx) + D\sin(kx), \tag{6.17}$$

where C and D are constants that define the amplitude and initial phase of the sinusoid (Appendix D). The fact that Equation 6.17 is a solution to Equation 6.15 can be verified by differentiating Equation 6.17 twice to obtain Equation 6.15.

The *boundary conditions* are defined by the fixed ends of the guitar string, which require that $f(x) = 0$ at both $x = 0$ and $x = L$. Consequently, for a guitar string of length L, the function $f(x)$ is zero at $x = 0$ and $x = L$, from which it follows that C must be zero.

The boundary conditions mean that there must be an integer number of half-wavelengths in L, that is, $n(\lambda/2) = L$. This implies that the wavenumber $k = 2\pi/\lambda$ can only adopt values that are multiples of π/L,

$$k_n \;\; = \;\; n \times (\pi/L) \tag{6.18}$$

for $n = 1, 2, \ldots$, so that Equation 6.17 becomes

$$f_n(x) \;\; = \;\; D_n \sin\left(\frac{n\pi}{L}x\right) \tag{6.19}$$

$$= \;\; D_n \sin(k_n x). \tag{6.20}$$

Note that if $n = 1$ then the wavelength is twice the length of the guitar string, if $n = 2$ then the wavelength is the length of the guitar string, and so on, as shown in Figure 2.3.

Evaluating $g(t)$. Substituting $G(t) = -\omega^2$ into Equation 6.14 and rearranging gives

$$\frac{d^2 g(t)}{dt^2} \;\; = \;\; -\omega^2 g(t). \tag{6.21}$$

The general solution to Equation 6.21 is a sinusoid with angular frequencies $\omega_n = v k_n$ for $n = 1, 2, \ldots$, so that

$$g_n(t) \;\; = \;\; M_n \sin(\omega_n t) + N_n \cos(\omega_n t), \tag{6.22}$$

where M_n and N_n are constants. For $n = 1$ the frequency is the *fundamental frequency* ω_0, for $n = 2$ the frequency is the *first harmonic* $\omega_1 = 2\omega_0$, and so on.

In summary, $f(x)$ defines the purely spatial sinusoid with a wavenumber of $k = 2\pi/\lambda$ rad/m, and $g(t)$ defines the purely temporal sinusoid with an angular frequency of $\omega = 2\pi/T$ rad/s. Thus, as stated in Equation 6.7, a standing wave is the product of independent spatial and temporal patterns (defined in Equations 6.20 and 6.22):

$$\Psi_n(x, t) \;\; = \;\; f_n(x)\, g_n(t) \tag{6.23}$$

$$= \;\; (D_n \sin k_n x)\,(M_n \sin \omega_n t + N_n \cos \omega_n t). \tag{6.24}$$

In the next section, we use the classical travelling and standing wave equations above as a template to obtain the Schrödinger wave equation. Specifically, we use the quantum equivalent of the sinusoid in Equation 6.6 to obtain a quantum travelling wave equation similar to the classical travelling wave in Equation 6.1, and we use the quantum equivalent of Equation 6.7 to obtain a quantum wave equation similar to the classical standing wave in Equation 6.12.

6.5. Schrödinger's Wave Equation

To obtain Schrödinger's wave equation, we need to define a few key quantities. Consider a particle with mass M travelling at velocity v, so that its momentum is $p = Mv$. As described in Section 4.2, a particle with momentum p can be expressed as a wave with de Broglie wavelength λ such that

$$p = \frac{h}{\lambda}. \tag{6.25}$$

As $\lambda = 2\pi/k$ where k is the wavenumber (Equation 5.5 on p75), $p = kh/(2\pi)$. The *reduced Planck's constant* is defined as $\hbar = h/(2\pi)$, so the wavenumber is

$$k = \frac{p}{\hbar}. \tag{6.26}$$

The energy of the particle is $E = h\nu$, with frequency $\nu = \omega/(2\pi)$, so

$$E = \hbar\omega, \tag{6.27}$$

and therefore the angular frequency is

$$\omega = \frac{E}{\hbar}. \tag{6.28}$$

Upon substituting Equations 6.26 and 6.28 into Equation 6.6, we have

$$\hat{\Psi}(x,t) = e^{i(px-Et)/\hbar}. \tag{6.29}$$

Just as the complex classical wave equation (Equation 6.5) can be obtained by differentiating a putative solution (Equation 6.6), so Schrödinger's (as yet unknown) wave equation can be obtained by differentiating the solution represented by Equation 6.29. Accordingly, our strategy is to 'work backwards' from Equation 6.29 to find the quantum wave that corresponds to the classical wave (Equation 6.1).

It is important to note that just as the fixed end points of the guitar string are responsible for the particular quantised frequencies, it is the fixed end points of the Schrödinger wave that are responsible for the particular quantised frequencies, which (in turn) determine the size of the energy quanta at each frequency.

The Time-Dependent Schrödinger Equation

To obtain the *time-dependent Schrödinger equation*, we begin by differentiating Equation 6.29 to evaluate the complex wave equation in Equation 6.5. To obtain the left-hand side of Equation 6.5, we differentiate Equation 6.29 with respect to x:

$$\frac{\partial \hat{\Psi}(x,t)}{\partial x} = (ip/\hbar)\hat{\Psi}(x,t). \tag{6.30}$$

Rearranging (and using $1/i = -i$) gives

$$-i\hbar\frac{\partial \hat{\Psi}(x,t)}{\partial x} = p\hat{\Psi}(x,t), \tag{6.31}$$

and then differentiating with respect to x again yields

$$-\hbar^2\frac{\partial^2 \hat{\Psi}(x,t)}{\partial x^2} = p^2\hat{\Psi}(x,t). \tag{6.32}$$

For a particle with mass M travelling at velocity v, its total energy E is the sum of its kinetic energy $Mv^2/2$ and its potential energy $V(x,t)$:

$$E = \tfrac{1}{2}Mv^2 + V(x,t). \tag{6.33}$$

Rewriting this in terms of the particle's momentum $p = Mv$, we have

$$E = p^2/(2M) + V(x,t). \tag{6.34}$$

Dividing Equation 6.32 by $2M$ gives

$$\frac{-\hbar^2}{2M}\frac{\partial^2 \hat{\Psi}(x,t)}{\partial x^2} = \frac{p^2}{2M}\hat{\Psi}(x,t), \tag{6.35}$$

where (from Equation 6.34) $p^2/(2M) = E - V(x,t)$; therefore

$$\frac{-\hbar^2}{2M}\frac{\partial^2 \hat{\Psi}(x,t)}{\partial x^2} + V(x,t)\hat{\Psi}(x,t) = E\hat{\Psi}(x,t). \tag{6.36}$$

Next, we differentiate Equation 6.29 with respect to t to obtain

$$\frac{\partial \hat{\Psi}(x,t)}{\partial t} = \frac{-iE}{\hbar}\hat{\Psi}(x,t), \tag{6.37}$$

which can be rearranged to yield

$$i\hbar\frac{\partial \hat{\Psi}(x,t)}{\partial t} = E\hat{\Psi}(x,t). \tag{6.38}$$

Equations 6.36 and 6.38 have the same right-hand side. Equating their left-hand sides yields the *time-dependent Schrödinger equation* (TDSE)

$$i\hbar \frac{\partial \hat{\Psi}(x,t)}{\partial t} = \left[\frac{-\hbar^2}{2M} \nabla^2 + V(x,t) \right] \hat{\Psi}(x,t), \qquad (6.39)$$

where ∇^2 denotes the second spatial derivative (∇ is called *del*, and ∇^2 is called the *Laplacian*). This is often written more succinctly as

$$i\hbar \dot{\hat{\Psi}}(x,t) = \hat{H}\hat{\Psi}(x,t), \qquad (6.40)$$

where $\dot{\hat{\Psi}}(x,t)$ is the temporal derivative and $\hat{H} = \frac{-\hbar^2}{2M} \nabla^2 + V(x,t)$ is known as the *Hamiltonian operator*.

The Time-Independent Schrödinger Equation

In general, if the potential energy is independent of time, so that $V(x,t) = V(x)$, then the wavefunction can be factorised. For example, if a particle is trapped in a sealed box then the energy of that particle is constant over time. This is analogous to the wave on a guitar string that is stationary at both ends, so it is effectively trapped; under ideal conditions, the guitar string has constant energy, so it will vibrate without attenuation forever. Under these conditions, the complex wavefunction can be separated into two complex functions (analogous to Equation 6.7 for the guitar string),

$$\hat{\Psi}(x,t) = \hat{f}(x)\,\hat{g}(t), \qquad (6.41)$$

where $\hat{f}(x)$ is a complex function of position only and $\hat{g}(t)$ is a complex function of time only. Thus, the TDSE (Equation 6.39) becomes

$$i\hbar\hat{f}(x)\frac{d\hat{g}(t)}{dt} = \hat{g}(t)\left[\frac{-\hbar^2}{2M}\frac{d^2}{dx^2} + V(x)\right]\hat{f}(x), \qquad (6.42)$$

and then dividing both sides by $\hat{f}(x)\hat{g}(t)$ yields

$$i\hbar \frac{1}{\hat{g}(t)}\frac{d\hat{g}(t)}{dt} = \frac{1}{\hat{f}(x)}\left[\frac{-\hbar^2}{2M}\frac{d^2}{dx^2} + V(x)\right]\hat{f}(x). \qquad (6.43)$$

As was the case for the classical Equation 6.12, we now have two functions of two different variables x and t, and the only way that two functions of different variables can be equal for all values of x and t is if they are equal to the same separation constant. For reasons that will become apparent, we will call that constant E.

Setting the left-hand side of Equation 6.43 equal to E and multiplying both sides by $\hat{g}(t)/i\hbar$ yields

$$\frac{d\hat{g}(t)}{dt} = -\frac{iE}{\hbar}\hat{g}(t), \qquad (6.44)$$

which depends only on time. If a function is proportional to its own derivative, as here, then that function must be exponential:

$$\hat{g}(t) = e^{\alpha t}, \qquad (6.45)$$

where α is a constant. We can verify that this is a solution because

$$\frac{d\hat{g}(t)}{dt} = \alpha\hat{g}(t). \qquad (6.46)$$

Comparison with Equation 6.44 implies that $\alpha = -iE/\hbar$, and therefore

$$\hat{g}(t) = e^{-iEt/\hbar}. \qquad (6.47)$$

Substituting this into Equation 6.41 yields a succinct form for the wavefunction (which will prove to be useful later),

$$\hat{\Psi}(x,t) = \hat{f}(x)e^{-iEt/\hbar}. \qquad (6.48)$$

Equation 6.47 can be used to show that E is energy, as follows. If we rewrite Equation 6.47 as

$$\hat{g}(t) = \cos(2\pi t E/h) - i\sin(2\pi t E/h) \qquad (6.49)$$

then it is apparent that $\hat{g}(t)$ is a sinusoidal function with frequency $\nu = E/h$. On the other hand, from the de Broglie equation (Equation 4.3) we know that the frequency of a particle is $\nu = E/h$ where E is the particle's energy. Therefore, the constant E in Equations 6.44–6.49 must, in fact, be *energy*.

As a reminder, we defined E to be the constant that the left- and right-hand sides of Equation 6.43 are both equal to. Therefore, we can set the right-hand side of Equation 6.43 equal to E and then multiply both sides by $\hat{f}(x)$ to obtain an equation that depends only on position, called the *time-independent Schrödinger equation*:

$$-\frac{\hbar^2}{2M}\frac{d^2\hat{f}(x)}{dx^2} + V(x)\hat{f}(x) = E\hat{f}(x). \qquad (6.50)$$

This can be used to find the probability distribution of particle positions in a container (Section 6.6) or in a hydrogen atom (Section 6.7).

6.6. Wavefunctions in a Box

Just as the waves on a guitar string can adopt only certain frequencies because the two ends of the string are held stationary, so the wavefunction of a particle confined to a box can adopt only certain standing wave frequencies because its two ends are effectively held stationary by the box walls. Because each wavefunction can adopt only discrete energy levels, which are proportional to frequency, it follows that each energy level depends on one standing wave frequency.

Consider a particle trapped inside a one-dimensional box of length L. The potential $V(x)$ is defined to be zero inside the box (i.e. for $x < L$) and infinite outside the box (i.e. for $x \geq L$). This box is usually known as an *infinite potential well*. The state of the particle is described by the TDSE (Equation 6.50) with $V(x) = 0$ (i.e. the particle is inside the box), so that Equation 6.50 becomes

$$-\frac{\hbar^2}{2M}\frac{d^2\hat{f}(x)}{dx^2} = E\hat{f}(x). \tag{6.51}$$

As $p = Mv$, we have $E = Mv^2/2 = p^2/(2M)$. From Equation 6.26, $p = \hbar k$, so we can rewrite the energy as

$$E = \frac{\hbar^2 k^2}{2M}. \tag{6.52}$$

If we substitute this into Equation 6.51 and cancel, we obtain

$$\frac{d^2\hat{f}(x)}{dx^2} = -k^2\hat{f}(x). \tag{6.53}$$

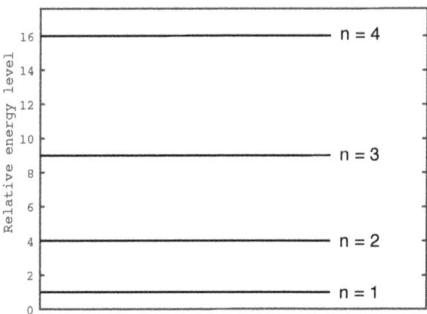

Figure 6.1: Energy levels in a one-dimensional container. Energy is measured in multiples of $\Delta E = \pi^2\hbar^2/(2ML^2)$ (Equation 6.57).

Note that this has a form identical to Equation 6.21, so we expect it to have similar solutions (Equations 6.18–6.20):

$$
\begin{aligned}
f_n(x) &= D_n \sin(k_n x) & (6.54) \\
&= D_n \sin(n\pi/L) & (6.55)
\end{aligned}
$$

where

$$
k_n = n\pi/L, \tag{6.56}
$$

such that each solution is associated with its own *quantum number n*. Substituting Equation 6.56 into Equation 6.52 yields

$$
E_n = n^2 \frac{\pi^2 \hbar^2}{2ML^2}, \tag{6.57}
$$

so energy levels increase in proportion to n^2, as shown in Figure 6.1. In classical physics, the lowest energy state of a particle is zero. In quantum mechanics, the lowest energy is the zero-point energy (p60) or *ground state*, which has $n = 1$, so $E_1 = \pi^2 \hbar^2/(2ML^2)$.

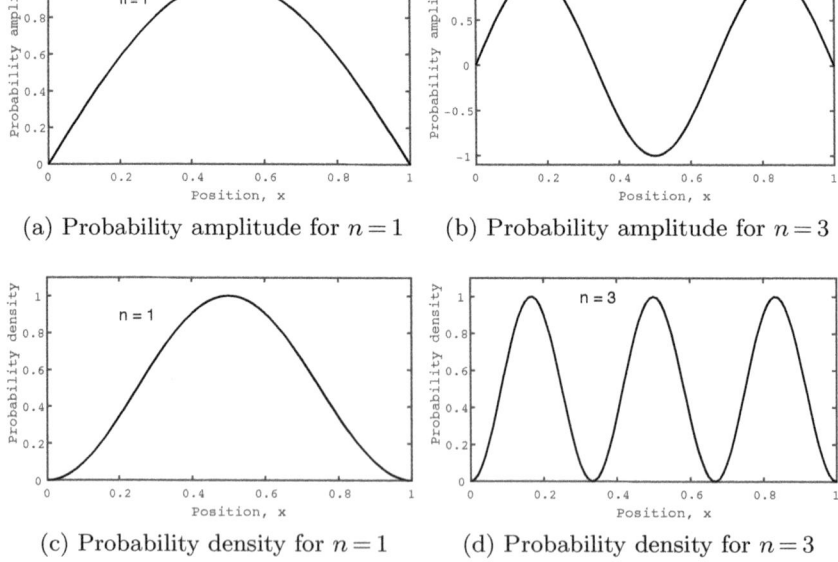

(a) Probability amplitude for $n = 1$ (b) Probability amplitude for $n = 3$

(c) Probability density for $n = 1$ (d) Probability density for $n = 3$

Figure 6.2: Position probability amplitude (a,b) and probability density function (c,d) (Equation 6.66) for a particle in a box.
(a) Probability amplitude $\Psi_1(x)$. (b) Probability amplitude $\Psi_3(x)$.
(c) Probability density $|\Psi_1(x)|^2$. (d) Probability density $|\Psi_3(x)|^2$.

Labelling each solution implicit in Equation 6.41 with a subscript n, we have the solutions

$$\hat{\Psi}_n(x,t) \;=\; f_n(x)\,\hat{g}_n(t) \tag{6.58}$$

where, from Equation 6.47,

$$\hat{g}_n(t) \;=\; e^{-iE_n t/\hbar}. \tag{6.59}$$

Substituting this and Equation 6.54 into Equation 6.58 yields

$$\hat{\Psi}_n(x,t) \;=\; [D_n \sin(k_n x)] \times [e^{-iE_n t/\hbar}]. \tag{6.60}$$

Interpreting Schrödinger's Equation. The complex wavefunction $\hat{\Psi}(x,t)$ (Equation 6.60) is a solution of Schrödinger's equation, and takes the place of the classical particle trajectory. In 1926, Max Born proved that the probability of finding the particle between positions x and $x + \Delta x$ at time t is

$$p(x,t)\,\Delta x \;=\; |\hat{\Psi}(x,t)|^2\,\Delta x, \tag{6.61}$$

which is known as the *Born rule*. For a particle trapped inside a container (from Equation 6.60),

$$
\begin{aligned}
|\hat{\Psi}_n(x,t)|^2 \Delta x \\
=\; & \hat{\Psi}_n^*(x,t)\,\hat{\Psi}_n(x,t)\,\Delta x \\
=\; & \left[D_n \sin(k_n x) \times e^{+iE_n t/\hbar}\right] \times \left[D_n \sin(k_n x) \times e^{-iE_n t/\hbar}\right] \Delta x \\
=\; & D_n^2\,\sin^2(k_n x)\,\Delta x. \tag{6.62}
\end{aligned}
$$

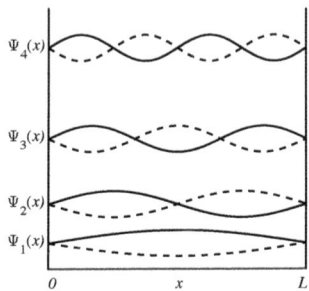

Figure 6.3: Probability amplitudes $\Psi_n(x)$ for quantum numbers $n = 1, 2, 3, 4$ for a particle confined to a box (Equation 6.67). Each amplitude oscillates around zero but is drawn at a height proportional to its energy ($E \propto n^2$). The wavefunctions $\Psi_1(x)$ and $\Psi_3(x)$ are the same as in Figure 6.2a and b (respectively).

Therefore, the probability associated with a complex wave does not depend on time, as noted on page 81.

For Equation 6.62 to be a probability density, its integral must evaluate to 1. Thus, setting

$$\int_{-\infty}^{\infty} |\hat{\Psi}_n(x,t)|^2 \, dx \;\; = \;\; |D_n|^2 \int_{0}^{L} \sin^2 k_n x \, dx \qquad (6.63)$$

$$= \;\; |D_n|^2 L/2 \qquad (6.64)$$

equal to 1 (following the reasoning given for Equation 5.43), we find that $D_n = \sqrt{2/L}$. Accordingly, if we substitute $D_n = \sqrt{2/L}$ into Equation 6.62 then the probability that a particle occupies a position within the interval Δx is

$$p(x) \, \Delta x \;\; = \;\; (2/L) \sin^2(k_n x) \; \Delta x. \qquad (6.65)$$

From Equation 6.56, $k_n = n\pi/L$, so

$$p(x) \, \Delta x \;\; = \;\; (2/L) \sin^2(n\pi x/L) \, \Delta x, \qquad (6.66)$$

which is plotted for $n=1$ and $n=3$ in Figure 6.2c and d, respectively.

We know from Equation 6.55 that the stationary wavefunction for position is sinusoidal, so it is unsurprising that Equation 6.66 defines each normal mode as a sinusoidal standing wave. Using the normalisation constant in Equation 6.64, we have

$$\Psi_n(x) \;\; = \;\; (2/L)^{1/2} \sin(n\pi x/L), \qquad (6.67)$$

where each value of n defines a probability amplitude function, as shown for $n=1$ and $n=3$ in Figure 6.2a and b, respectively, and in Figure 6.3 for $n = 1$–4.

6.7. Wavefunctions in a Hydrogen Atom

Just as a box containing electromagnetic waves only permits standing waves with certain wavelengths, so the positively charged nucleus of a hydrogen atom effectively acts as a spherical container, which only permits the standing waves of its orbiting electron to have certain wavelengths (see Figure 4.11). In the context of a hydrogen atom, the Schrödinger equation has solutions that define the probability of finding an electron at any point around the nucleus. Without going into the details of how the Schrödinger equation can be used to derive these solutions, the first two solutions (modes) correspond to the 1s and 2s orbital states. We will not worry about this notation, but for completeness, the '1' in '1s' refers to the *principal quantum number*, and 's' refers to the *angular quantum number* ('s' corresponds to an angular quantum number of zero).

The Schrödinger wavefunction of the 1s orbit is

$$\Psi(r) \;=\; \frac{2}{\sqrt{a_0}}\left(\frac{r}{a_0}\right)e^{-r/a_0}, \qquad (6.68)$$

as shown in Figure 6.4a, where the radius is plotted in units of Bohr radii a_0 (see Section 4.7). The probability of finding the electron within the volume $\Delta^3 r$ centred on the radius r is $|\Psi(r)|^2\,\Delta^3 r$, where $|\Psi(r)|^2$ is shown in Figure 6.4c. Bohr's calculation of a_0 matches the radius that gives the maximum value (probability) of the wavefunction for the 1s orbit. However, the information supplied by the Schrödinger equation in the form of wavefunctions is vastly richer than the scalar value of a_0 obtained with Bohr's model, described in Section 4.7.

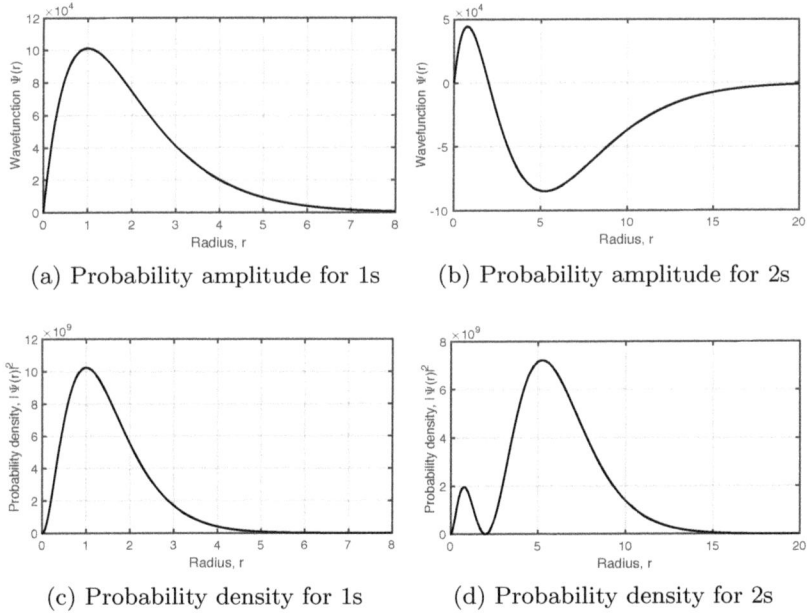

(a) Probability amplitude for 1s   (b) Probability amplitude for 2s

(c) Probability density for 1s   (d) Probability density for 2s

Figure 6.4: Probability amplitude (a,b; Equation 6.68) and probability density (c,d; Equation 6.69) of position for electron orbits 1s and 2s in a hydrogen atom.
(a) 1s probability amplitude $\Psi(r)$. (b) 2s probability amplitude $\Psi(r)$.
(c) 1s probability density $|\Psi(r)|^2$. (d) 2s probability density $|\Psi(r)|^2$.

The Schrödinger wavefunction of the 2s orbit is

$$\Psi(r) = \frac{2}{\sqrt{2a_0}} \left[1 - 0.5 \left(\frac{r}{a_0} \right) \right] \left(\frac{r}{a_0} \right) e^{-r/2a_0}, \qquad (6.69)$$

as shown in Figure 6.4b.

6.8. Summary

Schrödinger's wave equation simultaneously represents the most abstract and the most useful formulation of quantum mechanics. Inspired by de Broglie's idea that both matter and radiation can be represented as a wave, Schrödinger succeeded in finding an equation for that wave. But having an equation is irrelevant unless it can be used for practical purposes. For Schrödinger's equation, the mathematical machinery required for practical applications is not trivial. In this chapter, we have considered only the relatively simple application of calculating electron orbits, but there is much more machinery to be found in advanced texts.

Chapter 7

Quantum Interpretations

7.1. Introduction

Quantum mechanics, and especially Schrödinger's framework, describes everything in terms of waves. How, then, does quantum mechanics make contact with the macroscopic, too-solid world of objects?

The short answer is that no-one really knows. The longer answer is that the probabilities defined with such perfection by quantum mechanics yield singular events in the macroscopic world. The equations of quantum mechanics do not specify which events will occur, but they do specify the exact probability of each possible event. For example, we do not know the future location of a particle, but we do know (via Born's rule) the exact value of the probability that it will be in any given location. As we have seen, these probabilities are implicit in Schrödinger's wavefunction. Given such a wavefunction, what causes a particle to coalesce out of the miasma of quantum probabilities defined by that wavefunction?

Again, the short answer is that no-one really knows. The longer answer is that something disturbs the wavefunction, which somehow forces it to yield up one unique realisation out of the infinitude of possibilities it represents. This disturbance is usually referred to as the *collapse of the wavefunction*. A wavefunction collapse can be caused by something as simple as a photon materialising onto a photographic plate. Because this represents a kind of physical measurement, wavefunction collapse has come to be associated with the process of measuring the state of a physical system. The following sections provide a brief overview of the main interpretations of quantum mechanics.

7.2. The Copenhagen Interpretation

After a gestation period of a quarter of a century, quantum mechanics was born in 1926. With three fathers (Heisenberg, Schrödinger and Dirac) and many uncles, it was not long before the proud family began to worry about how to interpret the equations that defined the wondrous child they had created. Over the next few years, Niels Bohr and Werner

Heisenberg (among others), wrote a series of papers discussing the problem, resulting in the *Copenhagen interpretation.*

The Copenhagen interpretation assumes that the values of all physical quantities are undefined until they are measured, at which point the process of measurement induces the wavefunction to collapse. It is only this wavefunction collapse that forces each of the physical quantities measured to adopt a single value.

A key feature of the Copenhagen interpretation is that Schrödinger's wavefunction is not a property of the universe; rather, it is a descriptive tool that is useful inasmuch as it provides predictions regarding the state of the physical world, but the wavefunction itself cannot be said to correspond to anything physical. For all practical purposes, this seems to be the position adopted by Schrödinger. While developing ideas about his famous ambivalent cat, Schrödinger wrote in a letter to Einstein:

> *I am long past the state where I thought that one can consider the Ψ-function as somehow a direct description of reality.*

The Copenhagen interpretation is reminiscent of the outcome of a dispute between the Catholic church and Galileo (1564–1642). Galileo famously supported Copernicus' heliocentric theory that the Earth orbits the Sun. This view contradicted holy scripture and (more importantly) the Catholic church. Armed with Copernicus' theory, Galileo developed formulae for predicting the positions of heavenly bodies, which were widely used to aid navigation on the high seas. But after Galileo was threatened with torture for his heretical views, and (implicitly) the opportunity to be burned at the stake, he had to 'admit' that he had never supported the heliocentric theory. The compromise, or fudge, forced on Galileo was that his equations merely served the purpose of predicting the motions of the heavens; in the language of the time, Galileo's equations were merely a tool that could be employed to 'save the phenomena'. The difference is that, whereas Galileo was forced to downgrade his equations to mere 'hypothetical' descriptions of reality, Schrödinger voluntarily downgraded his wavefunction to a practical tool that could be employed to 'save the phenomena'.

A key outcome of the discussions between Bohr and Heisenberg was the *complementarity principle*, proposed by Bohr in 1928. This states that the values of certain pairs of physical quantities cannot both be known at the same time. For example, Heisenberg's uncertainty principle states that the position and momentum of a particle cannot both be known exactly. More generally, Bohr's complementarity principle implies that the entities in a double-slit experiment behave like waves or particles depending on the type of measurement imposed on the apparatus.

7.3. Objective Collapse Theories

Objective collapse theories treat the Schrödinger wavefunction as if it were a physical wave, which spontaneously collapses once the number of particles in an object becomes sufficiently large. A prominent version of this idea was published in 1986 by Ghirardi, Rimini and Weber, who proposed that each wavefunction collapses spontaneously on average every hundred million (10^8) years. This makes it sound as if a wavefunction collapse is a rare event. However, a 2 g cube of carbon (for example) contains about 10^{23} atoms, and if we associate a wavefunction with each atom then $10^{23}/10^8 = 10^{15}$ carbon atom wavefunctions in the cube will collapse each year, which comes to about 32 million wavefunction collapses per second. In other words, macroscopic objects comprise collections of wavefunctions that are collapsing pretty much all the time. Each wavefunction collapse acts like a measurement, causing a cascade of further collapses, which forces the entire object to remain as an object in the world of classical physics (i.e. a conventional object).

Note that the almost continuous spontaneous collapse of wavefunctions is no different from the inverse quantum Zeno effect discussed in Section 3.4. Whereas the inverse quantum Zeno effect relies on an external force to continuously measure the state of the system, spontaneous collapse involves the system behaving as if it continuously measures itself.

Because larger objects contain more particles, objective collapse depends on the size of the object under consideration. This, in turn, means that there should be an upper limit on the size of object that can exist in a superposition of quantum states. Thus, objective collapse theories are not merely abstract ideas — they make predictions that can be tested experimentally. If nothing else, spontaneous wavefunction collapse saves Schrödinger's cat from dangling in an undefined state, somewhere between life and death.

The philosophical problems generated by the mysterious wavefunction collapse clearly infuriated Schrödinger. As early as 1929, six years before he attempted to dispense with the whole idea by creating his famously half-dead cat, Schrödinger declared:

> If this damned quantum leaping is to remain, I regret having dealt with quantum theory at all.

7.4. Bohmian Mechanics

When de Broglie proposed in 1924 that particles, as well as light, behave like waves, he envisaged each particle being guided through space by a *pilot wave*. With the advent of Schrödinger's wave mechanics, it soon became clear that Schrödinger's wavefunction was a prime candidate for de Broglie's pilot wave. This general notion was developed by Bohm in 1952 into what is now called *Bohmian mechanics*.

An unusual aspect of Bohmian mechanics is that it is deterministic. However, removing the random element from quantum mechanics requires an additional equation for a particle's position. In practice, despite this apparent non-randomness, Bohmian mechanics makes the same predictions as standard quantum mechanics. Bohmian mechanics is historically important because it played a key role in the development of Bell's theorem (see Chapter 3).

7.5. The Many-Worlds Interpretation

As inscrutable as it is, the problems of wavefunction collapse seem trivial compared with the alternatives. For example, the *many-worlds interpretation*, which derives from the work of Hugh Everett, requires the creation of many entire new universes every time a measurement is made. Specifically, it is assumed that when a measurement with N possible outcomes is made, this causes the universe to split into N versions of itself. In each of these universes, life goes on as if a different one of the N outcomes had occurred.

An attractive feature of the many-worlds interpretation is that it is simple. However, this simplicity comes at a price, and it has been noted that the many-worlds interpretation is cheap on assumptions but expensive on universes.

7.6. The von Neumann–Wigner Interpretation

Because the human retina effectively measures the arrival of photons as efficiently as a photographic plate, it has been suggested that human consciousness is necessary for wavefunction collapse to occur[29]. In the limit, this seems to suggest that any event in the physical world exists as a mere potentiality until a sentient being observes it. At a single stroke, this solves the age-old riddle of whether or not a tree falling in an uninhabited forest makes a sound, or (as Einstein noted) whether the moon ceases to exist if no-one is looking at it. However, by analogy with the many-worlds interpretation, the von Neumann–Wigner interpretation is cheap on assumptions but expensive on sentient beings.

7.7. Summary

Plato proposed that our perceptions of reality are like the shadows cast on a wall; all we have are these shadows of a world that we can never see directly, cast by people and objects that we can never know with certainty. Even in this modern era, Plato's prosaic analogy serves as a cautionary tale for science, with a clear message that we should treat our observations of the world not as solid facts, but as provisional clues. As we have seen, when considered in the context of quantum mechanics, Plato's prescient analogy is more accurate than he could ever have imagined.

Chapter 8

A History of Quantum Mechanics

The discovery of simple and uniform principles, by which a great number of apparently heterogeneous phenomena are reduced to coherent and universal laws, must ever be allowed to be of considerable importance towards the improvement of the human intellect.
Young T, 1802.

Thus far, we have been concerned with the substance of quantum mechanics, rather than the particular historical order in which various discoveries were made. In fact, the order of presentation has been guided by increasing mathematical sophistication, so that ideas covered in one chapter act as a foundation for subsequent chapters.

Max Planck, a 42-year-old pillar of the scientific elite, had no intention of killing off classical physics. Despite this, he effectively hammered the first nail into the coffin of classical physics in 1900, and by 1926 it was under six feet of earth, with a headstone that could have said "Nice Try". By harnessing the tools introduced in Boltzmann's statistical mechanics, Planck was able to fit a function to the blackbody spectrum. However, the fit could be made exact only if Planck, in his famous act of desperation, was willing to sacrifice a central pillar of classical physics, namely that the energy of electromagnetic radiation can always be absorbed or emitted by matter in any continuous amount. In its place, Planck assumed that electromagnetic waves are absorbed or emitted only at discrete energy values. Specifically, the energy of a wave with frequency ν can change only in quanta of size $h\nu$, where h is Planck's constant. Planck was confident that his 'quantum trick' would soon be replaced by a more traditional, classical solution.

But within five years, the young Einstein had thrown the first handful of earth onto the coffin of classical physics. Prominent among the list of problems awaiting their turn to be solved by the self-anointed king of sciences was the *photoelectric effect* (explained below). Like blackbody radiation, the photoelectric effect was considered a tough problem, but one that would surely succumb to the scythe of classical physics.

It had been known for a long time that shining a light on metal somehow allows electric charge to flow away from the metal. And since the discovery of electrons by JJ Thomson in 1897, it was also known that this charge is carried by electrons, as if the light ejected electrons from the metal surface. In support of the idea that charge is carried by electrons, it was found that the amount of charge flowing away from an illuminated metal surface increases with the light intensity. However, it was also found that no electrons are ejected if the frequency of light falls below a critical value, even if the light is extremely intense; this is the so-called photoelectric effect. If, for example, radiation simply shook atoms in the metal until some electrons fell out then the amount of charge released should not depend on the frequency of light. But Einstein had read Planck's work, and the effect on him was profound:

> as if the ground had been pulled from under one, with no firm foundation to be seen anywhere.

In fact, the effect on Einstein was so profound that he proposed that not only is light absorbed or emitted by matter at discrete energy values or *quanta*, but light at a given frequency can only *exist* at discrete energy values. Einstein's *photoelectric equation* required the existence of packets of light, which eventually came to be known as *photons*. By 1909, Einstein had correctly predicted a *duality* between particles and waves, and that

> The next phase of development in theoretical physics will bring us a theory of light that can be interpreted as a kind of fusion of the wave and particle theories.

Despite a decade of resistance (from his fellow physicists) to the idea that light is quantised into photons, Einstein was awarded the Nobel prize for this work.

All of this was set against the backdrop of a physics community in search of the essence of matter and energy. At that time, positive and negative charges were thought to be distributed evenly throughout matter, which is why this was known as the *plum pudding* model. Then, in 1911, Ernest Rutherford fired alpha particles, known to be positively charged, at a thin sheet of gold foil. Most of the particles passed straight through the foil, but some were deflected to one side, and a few even reversed direction. In order to explain this, Rutherford postulated that the deflected alpha particles must be hitting something pretty solid, and he proposed that those somethings must be places where the positive charges of matter were concentrated. The result was Rutherford's *solar system* model of the atom, in which many negatively charged electrons orbit each positively charged nucleus, like planets around the sun.

But there was a problem: given that the nucleus is positively charged and that each electron is negatively charged, every electron ought to spiral down towards the nucleus. To salvage Rutherford's solar system atomic model, in 1913 Niels Bohr proposed that electrons can exist only in fixed orbits, which he called *stationary states*. Of course, Bohr knew of Planck's and Einstein's work on quantum energy, so it was natural to suppose that these stationary states were related to Planck's constant. Specifically, Bohr proposed that electron orbits have fixed, quantised *angular momenta* which exist only in multiples of $h/(2\pi)$ (the angular momentum of an electron with mass M travelling at speed v around a circular orbit of radius r is Mvr). As a result, quantised angular momenta implied corresponding quantised energy levels for electrons. In support of his model, Bohr confirmed the *Balmer formula* for the emission spectra of hydrogen atoms, which had hitherto been a purely empirical formula devised by Johann Balmer in 1885. Despite this success, Bohr's atom was considered a flawed mixture of classical and quantum ideas. Even so, his work acted as a spur to find a more complete quantum theory. In this way, Bohr contributed his own handful of earth to the final resting place of classical physics.

Over the next decade, the key ideas of Planck, Einstein and Bohr permeated the physics community. As is often the case, ideas that seemed barely credible to established physicists appeared quite natural to the newly minted physicists of the next generation. So it was that the young Prince Louis de Broglie proposed in 1924 that, just as light waves can behave as if they are particles, particles can behave as if they are waves, which exist only in discrete quanta. As de Broglie said:

> *When I conceived the first basic ideas of wave mechanics in 1923–24, I was guided by the aim to perform a real physical synthesis, valid for all particles, of the coexistence of the wave and of the corpuscular aspects that Einstein had introduced for photons in his theory of light quanta.*

But de Broglie did not just make a bold claim — he also quantified that claim with an equation. As a reminder, de Broglie's equation for a particle with momentum p is $p = h/\lambda$, where h is Planck's constant and λ is the wavelength. As de Broglie intended, his equation provided a quantum mechanical justification for Bohr's model.

Of course, an equation is of little use if it cannot be tested by experiment. In this case, de Broglie's equation predicted that electrons should diffract and interfere with each other to produce interference patterns similar to those observed in (light) double-slit experiments. Within three years that prediction had been verified, so there could be no doubt that matter, in the form of electrons, behaves as if it were being guided by what de Broglie called a *pilot wave*. This finally put to rest any argument regarding the relationship between light, energy and matter. In effect, de Broglie defined the universe in terms of just two natural kinds, waves and matter, with energy acting as the currency for converting between them. Not only was matter now considered to be a wave, but that wave was quantised by the same constant used to quantise light. Like Planck, Einstein and Bohr, de Broglie had cast his own handful of earth into the grave of classical physics.

The search for a coherent quantum theory was almost over. In 1925, the German *wunderkind* Werner Heisenberg effectively reinvented matrix algebra in the process of constructing *matrix mechanics*[15]. At that time, matrix algebra was not familiar to most physicists, which is probably why Heisenberg unwittingly reinvented it, and why his matrix mechanics was not widely adopted. In contrast, physicists were familiar with Maxwell's equations for electromagnetic waves, which may be why they found Erwin Schrödinger's *wave mechanics* much easier to digest.

In 1926, Schrödinger acted on a suggestion from Debye that if there is a de Broglie wave then there should be a wave equation to go with it, and the result is now known as the *Schrödinger equation*. This equation cannot be derived from classical physics, nor using any conventional process of deduction; it is essentially a product of Schrödinger's intellect. Later that year, Schrödinger showed that his wave mechanics and Heisenberg's matrix mechanics are equivalent.

In that remarkable year, Paul Dirac, having read a draft of Heisenberg's thesis in 1925, devised yet another framework for quantum mechanics, called *transformation theory*[9]. Happily, Dirac and Jordan were able to prove that Heisenberg's matrix mechanics and Schrödinger's wave mechanics are special cases of Dirac's transformation theory. Thus, by the end of 1926, there were three different, but equivalent, formulations of quantum mechanics. Three more handfuls of earth had now rendered classical physics almost invisible.

However, all three formulations suffered from an apparent anomaly, namely their prediction that if the momentum of a particle could be measured exactly then its position could be anywhere in the universe, and if the position of a particle could be measured exactly then its momentum could have any value. More precisely, all three formulations predicted that the product of the uncertainties in momentum and position cannot be less than a critical value, so if one is made smaller then the other must be made larger. This is the famous *Heisenberg uncertainty principle*, discovered in 1927.

Of course, no matter how elegant a theory is, and no matter how many physicists have developed equivalent formulations of that theory, it must eventually make contact with the everyday world of experience. Even though quantum mechanics was supremely good at predicting experimental results at the microscopic scale of particles, those predictions should always be compatible with experimental results at the macroscopic scale of everyday objects. This general idea was formalised by Bohr in 1923 as the *correspondence principle*, which states that the predictions of quantum mechanics must become indistinguishable from the predictions of classical physics at macroscopic scales. In other words, the microscopic physics of quantum mechanics must be consistent with the macroscopic world of classical physics.

Despite the enormous progress that had been made up to 1927, this was by no means the end of the story of quantum mechanics. In the years that followed, lively debates raged over the interpretation of quantum mechanics. Of these debates, one has a special status, not only because it took about thirty years to resolve, but because it represented the final handful of soil cast into the neglected grave of classical physics.

After several false starts, Albert Einstein and his colleagues Boris Podolsky and Nathan Rosen finally came up with a watertight proof that the predictions of quantum mechanics were incompatible with reality; more specifically, the predictions were incompatible with Einstein's own theory of special relativity, published in 1905. In essence, the famous EPR paper, published in 1935, predicted that two particles could be created in such a way that they share the same Schrödinger wavefunction, so that they are entangled. This means that if the state of one of the particles is measured then the other particle instantly adopts a state compatible with that of its partner, as if the particles can communicate with each other. Such instantaneous action can occur only if the particles communicate with each other at a speed that is effectively infinite. However, the instantaneous communication predicted by the EPR paper is incompatible with special relativity, which states that faster-than-light communication is impossible. The problem was that, for almost thirty years, no-one could come up with an experiment that could test the predictions of the EPR paper.

Then, in 1964, John Bell published his ingenious theorem, which showed how the idea of the EPR paper could be tested. Over the subsequent years, increasingly accurate experiments were carried out, and the increasingly convincing results yielded the same conclusion. Today, it is generally accepted that measuring the state of one particle not only determines the state of that particle but also determines the state of its entangled partner, as if they communicate with each other instantly. Ironically, what began life as a thought experiment designed to undermine the validity of quantum mechanics ultimately provided the means to remove all possible doubt of its truth.

Summary

1900 Planck proposes that light interacts with matter in discrete quanta.

1905 Einstein proposes that light exists only as quanta, which came to be known as photons.

1913 Bohr announces the solar system model of electron orbits and applies Planck's quantization idea to angular momentum, thus explaining the emission lines of the hydrogen spectrum.

1924 de Broglie proposes that matter behaves like a wave, and provides an equation relating momentum to the wavelength of matter waves.

1925 Heisenberg develops matrix mechanics in his doctoral thesis and publishes the first paper on quantum mechanics [15]. Schrödinger uses de Broglie's matter waves to develop wave mechanics, published in 1926.

1926 Dirac proves that the models of Heisenberg and Schrödinger and his own transformation mechanics are equivalent.

1927 Heisenberg's uncertainty principle is published. The existence of matter waves is confirmed for electrons.

1935 The EPR paper claims that quantum mechanics predicts particle entanglement, which implies faster-than-light communication, contradicting Einstein's special theory of relativity from 1905.

1935 Schrödinger's cat is born.

1964 Bell's theorem provides a recipe for testing the particle entanglement predicted by the EPR paper.

1972 Clauser provides the first experimental evidence that Bell's inequality is violated and, therefore, that Einstein's 'spooky action at distance' exists [13].

1976 Ingarden publishes the paper "Quantum Information Theory".

1982 Aspect confirms Clauser's result and closes the locality loophole.

1985 Deutsch describes the first universal quantum computer.

1994 Shor invents the first quantum computer algorithm.

2000 McFadden publishes the book *Quantum Evolution*.

2004 Quantum-state teleportation between calcium ions is demonstrated.

Further Reading

If I had to choose one book with a fine balance between words, diagrams and maths, it would have to be *Introduction to Quantum Mechanics* by AC Phillips. A close second would be McEvoy and Zarate's *Quantum Theory: A Graphic Guide*; despite its small size and cartoonish appearance, it contains a comprehensive and intuitive account of the development of quantum mechanics.

Al-Khalili, J (2019). *The Secret Of Quantum Physics: Einstein's Nightmare*. Part of the BBC Horizon series, available on YouTube, this episode examines the double-slit experiment in detail.
https://www.youtube.com/watch?v=f_4nYgrDJvc

Egan, G (1999). *Foundations 4: Quantum Mechanics*. Excellent informal primer of quantum mechanics (with equations). Available at
https://www.gregegan.net/FOUNDATIONS/04/found04.html

Elert, G (2020). *The Physics Hypertextbook*. Free online book with clear explanations and much historical background.
https://physics.info

Feynman, R (1964). *The Feynman Lectures on Physics*. These three volumes of lectures are widely recognised as the gold standard in physics books. Freely available at
https://www.feynmanlectures.caltech.edu/

Feynman, R (1964). *The Character of Physical Law - Part 6 Probability and Uncertainty*. Any lecture by Feynman is worth watching; this one, on the double-slit experiment, is a good primer for Chapter 5.
https://www.youtube.com/watch?v=aAgcqgDc-YM&list=
PLLzGzdSNup631MYeOpU9Hax6MBsTjdDas&index=6

Fitzpatrick, R (2019). *Introductory Quantum Mechanics*. Free online book with many excellent diagrams.
https://phys.libretexts.org/Bookshelves/Quantum_Mechanics/
Book%3A_Introductory_Quantum_Mechanics_(Fitzpatrick)

Greene, B (2004). *The Fabric of the Cosmos*. The clearest, and most thought-provoking, account of modern physics (without equations).

Lightman, A (2010). *The Discoveries: Great breakthroughs in 20th-century science.* Contains original text of key papers on physics and biology, each accompanied by a lucid commentary.

McEvoy, JP and Zarate, O (2007). *Quantum Theory: A Graphic Guide.* See above.

Melvyn, BM (2020). *Paul Dirac.* Discussion and biography of Dirac from BBC Radio 4 series In Our Time.
https://www.bbc.co.uk/programmes/m000fw0p

Nave, R (2017). *HyperPhysics.* Award-winning online account of physics from Georgia State University, originally designed for high school physics teachers, with fine sections on quantum mechanics.
http://hyperphysics.phy-astr.gsu.edu/hbase/quacon.html

Nelson, P (2017). *From Photon to Neuron.* A magnificent book, which is sadly not really about quantum mechanics; but Appendix B contains a lucid account of how Planck deduced the existence, and even the value, of (what is now known as) Planck's constant in 1899 (i.e. before he had discovered his own equation for blackbody radiation presented in December 1900). For more detail, see Stone (2013) below.

Phillips, AC (2003). *Introduction to Quantum Mechanics.* My favourite (see above).

Reich, H (2020). *Bell's Theorem: The Quantum Venn Diagram Paradox.* This Minutephysics video on polarisation is a good adjunct to Chapter 3.
https://www.youtube.com/watch?v=zcqZHYo7ONs&t=8s

Riley, KF (1974). *Mathematical Methods for the Physical Sciences: An Informal Treatment for Students of Physics and Engineering.* The best book on mathematics I have ever read.

Sanderson, G. *3blue1brown.* Any video lessons by Grant Sanderson's 3blue1brown are worth watching, especially those on quantum physics.
https://www.3blue1brown.com/
For Heisenberg's uncertainty principle:
https://www.youtube.com/watch?v=MBnnXbOM5S4&t=17s
For Bell's theorem:
https://www.youtube.com/watch?v=zcqZHYo7ONs

Stone, AD (2013). *Einstein and the Quantum.* A wonderful exposition of the early history of quantum mechanics.

van der Waerden, BL (1967). *Sources of Quantum Mechanics.* English translation of original papers published between 1916 and 1926. Begins with a forensic introductory section, which assumes substantial expertise. Published by Dover in 2009.

Appendix A

Glossary

agreement rate For a single photon that attempts to pass through filter A and then filter B, this is the percentage of occasions on which the photon passes through B given that it has passed through A. For a pair of photons, it is the percentage of occasions on which both photons pass through or are absorbed by their respective filters.

amplitude The amplitude of a wave is its maximum (e.g. time-varying) height.

angular frequency For a wave with a period of T seconds, the angular frequency is $\omega = 2\pi/T$ radians per second (rad/s). See **wavenumber**.

angular momentum Unless stated otherwise, this refers to the orbital angular momentum (as opposed to the spin momentum), which is the circular analogue of linear momentum. For a mass M moving at velocity v in a circular orbit with radius r, it is $p_{\text{angular}} = rMv$.

Balmer series Emission lines of hydrogen for which Balmer discovered a formula in 1885.

blackbody radiation The spectrum of electromagnetic frequencies emitted by a blackbody, approximated by the spectrum of radiation in an oven.

Bohr radius The radius of an electron orbit in a hydrogen atom in its ground state, $a_0 = 0.529 \times 10^{-10}$ m,

Boltzmann constant $k_B = 1.380649 \times 10^{-23}$ joules per kelvin (J/K).

Born rule The probability density of a particle being at a given position is the squared amplitude of the particle's wavefunction at that position.

bra-ket notation See Dirac notation.

complementarity Heisenberg's uncertainty principle states that certain conjugate or complementary pairs of variables (e.g. position–momentum and energy–time) cannot be measured exactly at the same time. Less formally, it is not possible to observe an entity as a wave and as a particle simultaneously.

complementary variables See complementarity.

conjugate variables See complementarity.

complex number A number consisting of two parts, a *real part* and an *imaginary part*; see Appendix D.

correspondence principle This states that the results of quantum mechanics must become identical to those of classical physics as the size of quantum effects becomes insignificant.

decoherence The disentangling of entangled particles, thought to be caused by a measurement process.

Dirac notation Dirac's bra-ket vector notation is used to represent quantum states, equivalent to Schrödinger's wavefunction.

electronvolt (eV) The charge on an electron is 1.602×10^{-19} C (coulombs). When accelerated by an electric potential of one volt, an electron has a kinetic energy of $1 \text{ eV} \approx 1.602 \times 10^{-19}$ J.

entangled If two particles are entangled then any measurement made on one particle affects the value of the same measurement made on the other particle.

entropy See Boltzmann's thermodynamic entropy or Shannon's information entropy.

expectation Often written $E[x]$, this is the mean value of a variable x.

Fourier transform Almost any function $f(x)$ can be expressed as a weighted sum of sinusoids, where each weight is a complex number that specifies the amplitude and phase of each sinusoidal frequency in $f(x)$. The set of weights constitute the Fourier transform of $f(x)$.

frequency Number of cycles per second, measured in units of s^{-1}, or hertz (Hz). Given that each cycle is 2π radians, a frequency of ν Hz equals an angular frequency of $\omega = 2\pi\nu$ rad/s.

ground state See zero-point energy.

Heisenberg's uncertainty principle See uncertainty principle.

height In this book, height refers to a wave's instantaneous magnitude.

hertz (Hz) A unit of frequency, corresponding to one complete cycle of 2π radians per second.

information entropy A measure of uncertainty. If x adopts values $x_1, \ldots, x_N$ with probabilities $p(x_1), \ldots, p(x_N)$ then the Shannon entropy of $p(x)$ is the average information associated with each value of x, which is $H(x) = \sum_i p(x_i) \log_2[1/p(x_i)]$ bits.

intensity The rate at which energy is delivered to a surface, measured in watts per square metre (W/m^2).

ionization energy The amount of energy required to remove an electron from an atom, measured in electronvolts. The ionization energy for hydrogen is 13.61 eV.

kelvin Temperature in kelvins (K); 273 K is the freezing point of water.

locality loophole A loophole which means that Bell's inequality could be violated without entanglement if the interval between measurements taken on entangled particles at x_1 and x_2 is no shorter than the time taken for light (and therefore information) to travel from x_1 to x_2.

matrix mechanics Heisenberg's formulation of quantum mechanics in terms of matrices. See Schrödinger's wave mechanics.

modulus Length of a vector z, e.g. for $z = (x, y)$ it is $|z| = \sqrt{x^2 + y^2}$. Also refers to the magnitude of a complex number (see Appendix D).

momentum In classical physics, this is just speed multiplied by mass, $p = Mv$. In quantum physics, momentum is quantised such that $p = h\nu$, where h is Planck's constant and ν is frequency.

normal mode A wave oscillating at a resonant frequency of a system (e.g. every harmonic of a guitar string is a normal mode).

observable A physical quantity, such as position or momentum.

orthogonal Perpendicular, at $90°$ to a reference axis.

phase velocity The speed $v = \omega/k$ at which a wave's peak travels, where k is the wavenumber and ω is the angular frequency.

photoelectric effect If a light is shone onto metal then electrons are released, but only if the light is above a certain frequency.

photon Frequency-dependent quantum of electromagnetic radiation.

Planck's constant $h \approx 6.626 \times 10^{-34}$ J/Hz, the quantum of energy required to increase the frequency of an electromagnetic wave by 1 Hz. The energy of a wave with frequency ν is $E = h\nu$.

probability density Probability per unit area (or volume), so that the probability over a small area ΔA centred on x is $P(x) \approx p(x)\Delta A$.

quantum The smallest possible change in a quantity (e.g. energy) at a given frequency.

quantum state A mathematical object that provides complete information on the probability of an observable (i.e. physical quantity such as position or momentum) adopting each value. Compare to **state**.

radian There are 2π radians in a circle, so 1 radian $= 360/(2\pi) \approx 57.3°$.

reduced Planck's constant $\hbar = h/(2\pi)$, where h is Planck's constant. Defines the quantum of angular momentum $M_e vr = n\hbar$ (in units of $\mathrm{kg\,m^2\,s^{-1}}$) for orbiting electrons (Equation 4.28).

rest mass The mass M_0 of a particle at rest. In contrast, special relativity states that the mass of a particle travelling at speed v is $M = M_0/\sqrt{1 - v^2/c^2}$. For a mass-less particle, such as a photon, its *relativistic mass* is derived from $E = Mc^2 = h\nu$ as $M = h\nu/c^2$.

Rybderg energy constant See ionization energy.

Rydberg constant (R) Relates two electron orbits n_1 and n_2 to the wavelength λ of the photon emitted when the electron falls from n_2 to n_1: $1/\lambda = R(1/n_1^2 - 1/n_2^2)\,\mathrm{m^{-1}}$, where $R = 1.097 \times 10^7\,\mathrm{m^{-1}}$.

Schrödinger equation There are two Schrödinger equations, the *time-independent Schrödinger equation* and the *time-dependent Schrödinger equation*. Both are differential equations, and their solutions are embodied in complex functions $\hat{\Psi}(x, y, z, t)$.

sinusoid Any wave defined by $y = C\cos\theta + D\sin\theta$ (see Section 5.4).

small-angle approximation For small angles α in radians, $\sin \alpha \approx \alpha$.

spatial frequency Spatial analogue of temporal frequency ν. For a wavelength λ, the spatial frequency is $\nu_S = 1/\lambda$ cycles/m (not to be confused with wavenumber).

spectral series The unique series of emission lines of a chemical element.

standard deviation For variable x with mean $\overline{x}$, its standard deviation is estimated as $\sigma = [(1/N) \sum_i^N (\overline{x} - x_i)^2]^{1/2}$.

state Unlike a quantum state, a state has a definite value. For example, the state of the observable mass has a definite value.

stationary wave See standing wave.

standing wave Also called a *stationary wave*, this is a wave with a fixed shape but an overall amplitude that varies over time, so the vertical distance between peaks and troughs changes over time, but the wave does not move left or right.

superposition The superposition of two waves is the linear sum of their values at every point. A quantum state can be represented as the superposition of other quantum states (Schrödinger equations).

theorem A mathematical statement that has been proved to be true.

thermodynamic entropy Boltzmann defined entropy in terms of the logarithm of the number Ω of equally probable states, $S = k_B \log \Omega$, where Boltzmann's constant is $k_B = 1.380649 \times 10^{-23}$ J/K.

uncertainty principle Given a pair of conjugate variables, such as the position x and momentum p of a particle, Heisenberg's inequality states that each variable can be measured up to an accuracy constrained by the condition $\sigma_x \sigma_p \geq \hbar/2$ J s, where σ_x is the standard deviation of x, σ_p is the standard deviation of p, and $\hbar$ is reduced Planck's constant.

wave equation Also known as the classical wave equation or linear wave equation, $\partial^2 \Psi(x,t)/\partial x^2 = (1/v^2)\partial^2 \Psi(x,t)/\partial t^2$; see Section 6.3.

wavefunction collapse When a wavefunction is measured, this forces it to collapse into a definite physical state. A measurement can be as simple as a photon wavefunction interacting with a photographic plate.

wavelength The distance λ between consecutive peaks of a wave.

wave mechanics Schrödinger's 1926 formulation of quantum mechanics in terms of waves.

wavenumber Given a sinusoidal wave such that one wavelength of λ metres is swept out for every complete revolution of 2π radians, the wavenumber is defined as $k = 2\pi/\lambda$ rad/m. Compare with angular frequency.

wave–particle duality This states that it is not possible to observe an entity as a wave and as a particle simultaneously. See complementarity.

zero-point energy The lowest energy level of a quantum system. Heisenberg's uncertainty principle guarantees that the zero-point energy has a fixed non-zero value.

Appendix B

Mathematical Symbols

Estimated values and complex variables have a hat symbol; for example, $\hat{\Psi}$ is a complex wavefunction, whereas Ψ is a (real) wavefunction.

$|A|$ a pair of vertical bars denotes a) absolute value, b) modulus (vector length), or c) magnitude of a complex number (see Glossary and Appendix D).

A amplitude, or magnitude or modulus, of a wavefunction.

a_0 Bohr radius, 0.529×10^{-10} m.

δ (delta) Kronecker or Dirac delta function.

C coulomb, unit of electric charge.

c speed of light, $299{,}792{,}458$ m/s $\approx 3 \times 10^8$ m/s.

E expectation or mean.

E energy (in units of joules, J).

E_{R} Rydberg energy constant, 13.6 eV or 2.2×10^{-18} J.

e the constant $2.71828\ldots$; also the electron charge, $\approx 1.60 \times 10^{-19}$ C.

eV electronvolt, unit of energy, $\approx 1.60 \times 10^{-19}$ J.

h Planck's constant, $\approx 6.626 \times 10^{-34}$ J/Hz.

$\hbar$ reduced Planck's constant, $h/(2\pi) = 1.055 \times 10^{-34}$ J s.

H Shannon information or entropy (measured in bits).

Hz hertz, unit of frequency, equivalent to s^{-1}.

i unit imaginary number, $i^2 = -1$.

J joule, unit of energy.

K kelvin, unit of temperature; 273 K is the freezing point of water.

k wavenumber, $k = 2\pi/\lambda = p/\hbar$ rad/m.

Mathematical Symbols

k_B Boltzmann's constant, 1.380649×10^{-23} J/K.

m metre, unit of length.

M mass (in units of kilograms, kg).

∇ (nabla) operator called del; ∇^2 represents the second derivative, called the Laplacian.

p momentum, classically mass $\times$ speed (in units of kg m/s).

nm nanometre, unit of length equal to 10^{-9} m.

Ψ (psi) a real-valued wavefunction.

$\hat{\Psi}$ (psi) a complex wavefunction.

Q common-sense agreement rate between the behaviour of one photon with respect to two filters, or between the behaviour of two photons with respect to their corresponding filters.

q quantum mechanical agreement rate between the behaviour of one photon with respect to two filters, or between the behaviour of two photons with respective to their corresponding filters.

R Rydberg constant, 1.097×10^7 m^{-1}.

r radius; also modulus of a complex number, e.g. $r = |z| = \sqrt{x^2 + y^2}$.

s second, unit of time.

σ (sigma) standard deviation.

t time (in units of seconds).

T temperature; or time for an oscillator to complete one cycle of 2π radians (period).

θ (theta) angle; phase of a wave.

ω (omega) angular frequency, $\omega = 2\pi\nu$ rad/s.

v velocity (in units of metres per second, m/s).

ν (nu) frequency (in units of cycles per second, s^{-1} or Hz).

ϕ general optic angle.

φ optic half-angle of bright region of diffraction envelope.

x position along the x-axis; also distance from slit.

y position along the y-axis.

z complex number, $z = x + iy$; also position along the z-axis.

z^* complex conjugate, $z^* = x - iy$, of the complex number $z = x + iy$.

Appendix C

The Boltzmann Distribution

The Boltzmann Distribution and Energy Barriers

Consider a sealed box that contains two bowl-shaped depressions at different heights, as depicted in Figure C.1. In the box there is a gas, which settles into the two bowls under the force of gravity. The box is well insulated, so that the total amount of energy in the box remains constant. Even though the gas molecules are free to move in any direction and at any speed, the average speed, and therefore the average kinetic energy, of each molecule depends only on the temperature within the box. Here, we will be concerned only with the height of each molecule, because height determines the potential energy of the molecule. Using this box of gas, we can answer a vital question of statistical mechanics: what is the probability that a randomly chosen molecule of gas has potential energy E? Specifically, what is the distribution of energies of gas molecules within the box?

Potential energy is proportional to height h, so molecules that have settled into the bowl b_1, which is at height h_1, have potential energy $E_1 = Gh_1$, where G is the gravitational constant. Similarly, molecules that have settled into the bowl b_2, which is at height h_2, have potential energy $E_2 = Gh_2$. Note that the barrier between the bowls prevents the molecules from simply all flowing down into bowl b_1. Gas molecules have a range of different speeds and directions, so by chance some molecules in b_2 will hop over the barrier into the lower bowl b_1, denoted by $b_2 \to b_1$. Similarly, some molecules in b_1 will hop over the barrier into the higher bowl b_2.

Eventually, the system settles into a stable equilibrium such that the rate of $b_2 \to b_1$ movement is exactly matched by the rate of $b_1 \to b_2$ movement. From this, we can deduce the distribution of potential energies of molecules within the box.

First, the probability that any molecule in b_1 will hop from b_1 to b_2 is the same for all molecules in b_1. Therefore, the rate dn_1/dt of $b_1 \to b_2$

123

transitions must be proportional to the number n_1 of molecules in b_1:

$$dn_1/dt \quad \propto \quad n_1. \tag{C.1}$$

Second, we assume that the rate at which molecules hop from b_1 to b_2 is related to the height of the barrier. For molecules in b_1, the effective barrier height is the difference between E_1 and E_a, which is

$$\Delta E_1 \quad = \quad E_a - E_1. \tag{C.2}$$

Specifically, we assume that dn_1/dt is proportional to the value of a function $f(\Delta E_1)$, so that

$$dn_1/dt \quad \propto \quad f(\Delta E_1). \tag{C.3}$$

Finding the correct form for this function f is our main objective.

Putting Equations C.1 and C.3 together yields

$$dn_1/dt \quad = \quad kn_1 f(\Delta E_1), \tag{C.4}$$

where k is a constant. Similarly, the barrier height for molecules in b_2 is

$$\Delta E_2 \quad = \quad E_a - E_2, \tag{C.5}$$

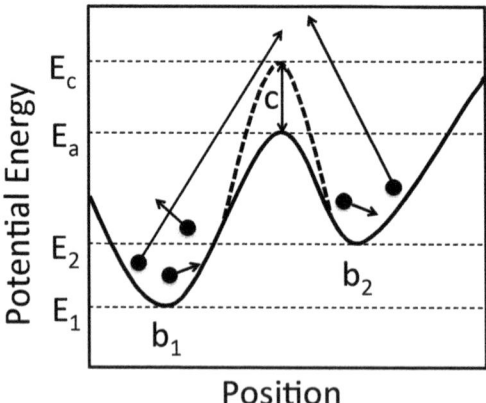

Figure C.1: Cross-section of a sealed box of gas with an inner contoured surface, where the potential energy E of each molecule is proportional to its height. The temperature is sufficiently low that molecules accumulate in the bowl-shaped depressions b_1 and b_2. The random motions of the molecules ensure a constant flow of gas from b_1 to b_2 and vice versa. After some time, the box reaches a state of equilibrium such that the $b_1 \to b_2$ flow equals the $b_2 \to b_1$ flow. The objective is to find a function $f(E)$ that defines the probability $p(E)$ of a particle having energy E.

and the rate dn_2/dt of $b_2 \to b_1$ transitions is

$$dn_2/dt \quad = \quad kn_2 f(\Delta E_2). \tag{C.6}$$

Crucially, at equilibrium, the rates in both directions are the same:

$$dn_1/dt \quad = \quad dn_2/dt, \tag{C.7}$$

so that

$$n_1 f(\Delta E_1) \quad = \quad n_2 f(\Delta E_2). \tag{C.8}$$

In essence, this states that if the number n_1 of molecules in b_1 is large (as seems likely) then $f(\Delta E_1)$ will be small, and vice versa. As we shall soon discover, the function $f(\Delta E_1)$ is proportional to the probability that a molecule hops from b_1 to b_2. This being so, we can see that a large number of molecules in b_1 combined with a small probability of a $b_1 \to b_2$ hop can equal a small number of molecules in b_2 combined with a large probability of a $b_2 \to b_1$ hop. To find a form for $f(\Delta E_1)$, we proceed as follows.

Suppose the barrier is raised from E_a to $E_c = E_a + c$. Now the rate of change between the bowls becomes

$$n_1 f(\Delta E_1 + c) \quad = \quad n_2 f(\Delta E_2 + c). \tag{C.9}$$

Dividing Equation C.8 by Equation C.9 gives the ratio of the rates before and after raising the barrier:

$$\frac{f(\Delta E_1)}{f(\Delta E_1 + c)} \quad = \quad \frac{f(\Delta E_2)}{f(\Delta E_2 + c)}. \tag{C.10}$$

This can be solved using Lagrange multipliers, or we could just guess a form for the function f that satisfies Equation C.10. Suppose we make a guess that $f(x) = e^{-x}$, so that we have

$$\frac{e^{-\Delta E_1}}{e^{-\Delta E_1 - c}} \quad = \quad \frac{e^{-\Delta E_2}}{e^{-\Delta E_2 - c}}. \tag{C.11}$$

Expanding the denominators yields

$$\frac{e^{-\Delta E_1}}{e^{-\Delta E_1} \times e^{-c}} \quad = \quad \frac{e^{-\Delta E_2}}{e^{-\Delta E_2} \times e^{-c}}, \tag{C.12}$$

which results in $e^c = e^c$. The fact that this identity is obtained confirms that our guess for f was a good one. In fact, the only function that satisfies Equation C.10 is an exponential function.

C The Boltzmann Distribution

From Equation C.8, the ratio of the numbers of molecules in the two bowls is

$$\frac{n_1}{n_2} = \frac{f(\Delta E_2)}{f(\Delta E_1)}. \tag{C.13}$$

Now that we know $f(\Delta E) = e^{-\Delta E}$, it follows that

$$\frac{n_1}{n_2} = \frac{e^{-\Delta E_2}}{e^{-\Delta E_1}} \tag{C.14}$$

$$= e^{\Delta E_1 - \Delta E_2}, \tag{C.15}$$

where

$$\Delta E_1 - \Delta E_2 = (E_a - E_1) - (E_a - E_2) \tag{C.16}$$
$$= E_2 - E_1. \tag{C.17}$$

Therefore

$$\frac{n_1}{n_2} = e^{E_2 - E_1} \tag{C.18}$$

$$= \frac{e^{-E_1}}{e^{-E_2}} \tag{C.19}$$

Thus, in general, the number n of molecules in a bowl at a height where the average potential energy of every molecule is E decreases as an exponential function of E:

$$n \propto e^{-E}. \tag{C.20}$$

The probability that a randomly chosen molecule has energy E_1 is

$$p(E_1) = \frac{n_1}{n_1 + n_2}, \tag{C.21}$$

and this, together with Equation C.20, gives

$$p(E_1) = \frac{e^{-E_1}}{e^{-E_1} + e^{-E_2}}. \tag{C.22}$$

In practice, $p(E)$ depends on the temperature T and on Boltzmann's constant k_B, which yields the Boltzmann distribution

$$p(E) = \frac{1}{Z} e^{-E/(k_B T)}, \tag{C.23}$$

where the constant Z is known as the *partition function* and ensures that the probabilities sum to 1:

$$Z = \sum_{j=1}^{N} e^{-E_j/(k_B T)}, \qquad (C.24)$$

where $N = 2$ in this example. Even though we have defined this problem in terms of energy, expressed using the height of molecules, height is really a proxy for the state of any particle, where different states have different energies. For example, state can refer to the frequency of a photon, or to the orbit occupied by an electron in an atom.

A Galilean Derivation of the Boltzmann Distribution

The following derivation of the Boltzmann distribution is analogous to Galileo's purely logical proof that all objects must fall at the same rate. We begin by making the basic assumption that the collection of all objects with energy E make up a proportion $f(E)$ of all objects, such that, aside from a few constants, the proportion $f(E)$ of all objects with energy E depends only on E. The objective is to find a form for the function $f(E)$ that is consistent with this assumption.

If a particle is chosen at random then the probability that it has energy E_1 is the proportion $f(E_1)$ of particles with energy E_1:

$$p(E_1) = f(E_1). \qquad (C.25)$$

Similarly, the probability that another particle chosen at random has energy E_2 is the proportion $f(E_2)$ of particles with energy E_2:

$$p(E_2) = f(E_2). \qquad (C.26)$$

The independence between particles means that if we were to choose two particles at random then the joint probability $p(E_1, E_2)$ that one particle has energy E_1 and the other particle has energy E_2 is just the product of their individual probabilities:

$$p(E_1, E_2) = p(E_1)p(E_2) \qquad (C.27)$$
$$= f(E_1)f(E_2). \qquad (C.28)$$

Next comes the crucial step. We treat all pairs of particles as single entities, called particle-pairs. Therefore, the probability $p(E_1, E_2)$ of choosing the particle-pair comprising two individual particles with energies E_1 and E_2 is the same as the probability $p(E_3)$ of choosing a third particle with energy $E_3 = E_1 + E_2$; that is,

$$p(E_3) \;=\; p(E_1, E_2). \qquad\qquad \text{(C.29)}$$

From Equation C.28, $p(E_1, E_2) = f(E_1)f(E_2)$, so that

$$p(E_3) \;=\; f(E_1)f(E_2). \qquad\qquad \text{(C.30)}$$

But the probability that this third particle has energy E_3 is just the proportion of particles with energy E_3, which is $f(E_3)$, so we have

$$p(E_3) \;=\; f(E_3) \qquad\qquad \text{(C.31)}$$
$$\;=\; f(E_1 + E_2). \qquad\qquad \text{(C.32)}$$

If we now treat f as a function then Equations C.30 and C.32 imply that f must satisfy

$$f(E_1)f(E_2) \;=\; f(E_1 + E_2). \qquad\qquad \text{(C.33)}$$

This equation can only be true if the proportion $f(E)$ of particles with energy E is an exponential function of energy,

$$f(E) \;\propto\; e^{-E}. \qquad\qquad \text{(C.34)}$$

We can confirm that the exponential function in Equation C.34 satisfies the constraint in Equation C.33 by substituting Equation C.34 in C.33:

$$e^{-E_1} \times e^{-E_2} \;=\; e^{-(E_1+E_2)}. \qquad\qquad \text{(C.35)}$$

Formally, Equation C.33 is also satisfied by the function form in Equation C.34 with a positive exponent. However, any solution with a positive exponent can be ruled out because it would imply that the probability of a state with energy E increases without limit as E increases. Using Equations C.25 and C.34, we can write

$$p(E) \;\propto\; e^{-E}. \qquad\qquad \text{(C.36)}$$

This is consistent with Equation C.22, so the same line of reasoning can be used here to obtain the Boltzmann distribution. Notice that Equation C.36 was derived via logical reasoning only, whereas the dependence on temperature and the value of Boltzmann's constant in Equation C.23 must be discovered through experiment.

Appendix D

Complex Numbers

A complex number consists of two parts, a *real part* and an *imaginary part*. The real part is a conventional real number, like 3 or 4.2 or π. The imaginary part is a real-number multiple of the *unit imaginary number i*, where i is the square root of minus one, so that $i^2 = -1$ and $1/i = -1$. A complex number $\hat{z}$ and its *complex conjugate $\hat{z}^*$* are usually written as

$$\hat{z} = x + iy \quad \text{and} \quad \hat{z}^* = x - iy, \tag{D.1}$$

where x and y are both real numbers; $x = \text{Re}(\hat{z})$ is the real part of $\hat{z}$, and $iy = i \times \text{Im}(\hat{z})$ is a *pure imaginary number*, equal to y lots of the unit imaginary number i. For consistency with notation in the rest of the book, we write complex numbers with a hat symbol here.

A complex number can be represented as a point on the *complex plane*, also known as an *Argand diagram* (Figure D.1). The real part is represented on the horizontal axis, and the imaginary part is represented on the vertical axis, so that z is represented by the *Cartesian coordinates* (x, y). It can also be represented as a vector in *polar coordinates* (r, θ). The length of this vector is given by the *absolute value*, *modulus* or

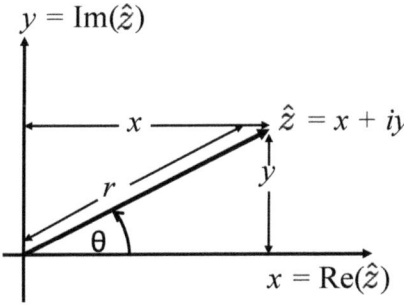

Figure D.1: Argand diagram representation of a complex number $\hat{z}$.

magnitude of the complex number $\hat{z}$,

$$r \;=\; |\hat{z}| \;=\; \sqrt{x^2 + y^2}. \tag{D.2}$$

The angle θ between the vector (x, y) and the x-axis is the *phase*, and its tangent is the ratio of the imaginary part to the real part of $\hat{z}$, so that $\tan\theta = y/x$. The real and imaginary parts of $\hat{z}$ can be expressed as $x = r\cos\theta$ and $y = r\sin\theta$. Substituting these into Equation D.1 yields

$$\hat{z} \;=\; r(\cos\theta + i\sin\theta). \tag{D.3}$$

Euler's theorem states that $e^{i\theta} = \cos\theta + i\sin\theta$, so Equation D.3 becomes

$$\hat{z} \;=\; re^{i\theta}. \tag{D.4}$$

The amplitude and initial phases of a complex sinusoidal wave $\hat{\Psi}(\theta)$ can be defined by a complex constant $\hat{A} = C + iD$, so that

$$\hat{\Psi}(\theta) \;=\; \hat{A}e^{i\theta} \;=\; (C + iD)(\cos\theta + i\sin\theta). \tag{D.5}$$

This can be decomposed into real and imaginary component waves (Figure 5.5),

$$\hat{\Psi}(\theta) \;=\; (C\cos\theta - D\sin\theta) + i(C\sin\theta + D\cos\theta). \tag{D.6}$$

The amplitude r of this wave is the modulus of $\hat{\Psi}(\theta) = \hat{A}e^{i\theta}$. As the modulus of a product of complex numbers $\hat{z}_1$ and $\hat{z}_2$ is $|\hat{z}_1\hat{z}_2| = |\hat{z}_1||\hat{z}_2|$,

$$r \;=\; |\hat{\Psi}(\theta)| \;=\; |\hat{A}e^{i\theta}| \;=\; |\hat{A}||e^{i\theta}|. \tag{D.7}$$

Now the amplitude can be obtained from the squared modulus $r^2 = |\hat{A}|^2|e^{i\theta}|^2$, because $|e^{i\theta}|^2 = e^{i\theta}e^{-i\theta} = 1$, which means that

$$r^2 \;=\; |\hat{A}|^2 \;=\; \hat{A}\hat{A}^* \;=\; (C + iD)(C - iD) \;=\; C^2 + D^2, \tag{D.8}$$

and therefore the amplitude is $r = \sqrt{C^2 + D^2}$. The initial phases of the real and imaginary parts of the sinusoid are $\theta_{\text{Real}} = \arctan(-D/C)$ and $\theta_{\text{Imag}} = \arctan(C/D)$, so the difference between the real and imaginary angles is always $\pi/2$, as we would expect from Equation D.3.

If $\hat{\Psi}(\theta)$ represents a complex travelling wave (where θ changes over time) then it describes a corkscrew-shaped trajectory with a constant height (Figure 5.5). In contrast, the height of a real wave varies with θ (e.g. $\text{Re}(\hat{\Psi}(\theta)) = C\cos\theta - D\sin\theta$ in Equation D.6, Figure 2.1). Complex amplitudes are useful for setting the phase and amplitude of Fourier components (Appendix F). If $\hat{\Psi}(\theta)$ is a Schrödinger wavefunction then the normalisation requirement (Equation 5.42) means that the value of $|\hat{A}|$ is arbitrary. For a discursive account, see Egan (Further Reading).

Appendix E

Pioneers of Quantum Mechanics

Bohr, Niels (1885–1962) In 1913, Bohr (with Rutherford) proposed that electrons in atoms have *stationary orbits* and that the angular momentum of adjacent orbits can adopt values only in increments of $h/(2\pi)$, where h is Planck's constant.

Born, Max (1882–1970) Born and his former student Jordan used vector–matrix notation to formalise a draft paper by Heisenberg and so develop *matrix mechanics*. He proposed the Born rule, which states that the probability density of a particle being at a given position is the square of the magnitude (probability amplitude) of the particle's wavefunction.

de Broglie, Louis-Victor-Pierre-Raymond (1892–1981) In 1924, de Broglie proposed in his doctoral thesis that all matter behaves like waves. Specifically, he proposed that an object with momentum p has a wavelength of $\lambda = h/p$, where h is Planck's constant.

Dirac, Paul (1902–1984) In 1926, Dirac proved that Heisenberg's matrix mechanics and Schrödinger's wave mechanics are special cases of Dirac's own *transformation theory*.

Einstein, Albert (1879–1955) In 1905, Einstein explained the photoelectric effect by assuming that light energy is quantised into corpuscles or photons. Einstein proposed that light itself is quantised.

Heisenberg, Werner (1901–1976) In 1925, Heisenberg, Born and Jordan published a paper on *matrix mechanics* that represents the first formal account of quantum mechanics. In 1927, Heisenberg proposed what came to be known as Heisenberg's uncertainty principle.

Pauli, Wolfgang (1900–1958) Originator of Pauli's exclusion principle, which states that only one electron can occupy a given state in an atom.

Planck, Max (1858–1947) In 1900, Planck came up with an equation that fitted the blackbody radiation spectrum, by assuming that when radiation is emitted or absorbed by matter, its energy is proportional to its frequency, where the constant of proportionality is now known as Planck's constant. Thus, the energy of light with frequency ν emitted or absorbed by matter is $E = h\nu$, where h is Planck's constant. Planck proposed that light can only be emitted or absorbed by matter at certain frequencies, but it was Einstein who proposed in 1905 that light itself is quantised.

Schrödinger, Erwin (1887–1961) In 1926, Schrödinger published a paper that contained his equation $i\hbar\dot{\Psi} = \hat{H}\Psi$, where Ψ is the wavefunction, $\dot{\Psi} = \partial\Psi/\partial t$, $\hbar = h/(2\pi)$, h is Planck's constant, and $H = \frac{-\hbar^2}{2M}\nabla^2 + V(x,t)$ is known as the Hamiltonian operator.

Appendix F

Fourier Optics and Heisenberg

Heisenberg's uncertainty principle depends critically on *Heisenberg's inequality*, which follows from a theorem that applies to any pair of conjugate variables. Heisenberg's inequality has a particularly intuitive interpretation in the context of *Fourier optics*, in which an image produced by a simple camera is the *Fourier transform* of the shape of the camera's aperture. This general result can be applied to a relevant example of Fourier optics in which the aperture is a vertical slit.

Fourier Analysis

Joseph Fourier (1768–1830) was one of several scientists involved in the development of Fourier analysis. This proves that, for all practical purposes, any reasonable function can be constructed from a superposition of sinusoids with different phases and amplitudes. But the really clever part is that Fourier analysis provides a method for finding precisely which amplitudes and phases are required to (re)construct such a function. Thus, given a function Ψ and a set of sinusoids with an infinite number of different frequencies, Fourier analysis provides the exact phase and amplitude of the sinusoid at each frequency such that when these sinusoids are added together, the result is the function Ψ, as shown in Figure F.1. For an excellent brief overview of Fourier analysis, see parts 1 and 2 of Brian Douglas's video at https://www.youtube.com/watch?v=1JnayXHhjlg.

Diffraction and Matter Waves

A vertical slit is the spatial equivalent of a temporal pulse, like the sound of a hand clap. And just as a temporal pulse can be represented as a superposition of sinusoids (Fourier components) of different temporal frequencies and wavelengths, so a spatial pulse can be represented as a superposition of Fourier components with different spatial frequencies and wavelengths. In the case of a slit, each Fourier component is uniquely associated with a particular direction ϕ, expressed either as a wavenumber k_ϕ or as a spatial frequency (equal to $k_\phi/(2\pi)$).

Recall the physical apparatus shown in Figure 4.4 of Section 4.5. All particles from a distant light source enter the vertical slit with the same horizontal momentum p and with zero momentum p_y along the vertical axis. However, diffraction at the slit causes particles to exit the slit at different angles. Because the total amount of each particle's momentum remains constant, a particle that enters the slit with momentum p and leaves the slit at angle ϕ has a vertical momentum of

$$p_y(\phi) \quad = \quad p\sin\phi. \tag{F.1}$$

To harness Fourier analysis, we need to express all variables in terms of waves. This can be achieved by using de Broglie's formula $p = \hbar k$ to express momentum in terms of the wavenumber k, so that

$$p_y(\phi) \quad = \quad \hbar k \sin\phi. \tag{F.2}$$

Again from de Broglie's formula, we can treat $k\sin\phi$ as a wavenumber

$$k_\phi \quad = \quad k\sin\phi, \tag{F.3}$$

so that Equation F.2 becomes

$$p_y(\phi) \quad = \quad \hbar k_\phi. \tag{F.4}$$

Crucially, this provides a link between the vertical momentum $p_y(\phi)$ of a putative particle and the wavenumber k_ϕ of a wave. The key idea is that each direction ϕ has a unique wavenumber k_ϕ associated with it, and each position y within the slit contributes the same amount $\hat{\Psi}(k_\phi)$

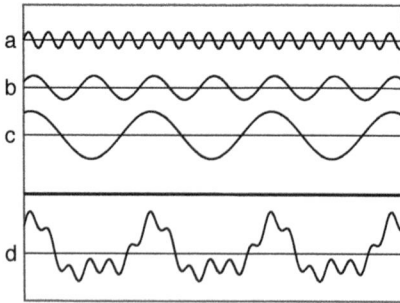

Figure F.1: Fourier analysis. The complicated curve in d is the sum of corresponding points in the sinusoidal curves in a–c. Fourier analysis decomposes the curve in d into sinusoidal functions with the unique set of amplitudes and phases shown in a–c.

of that wavenumber to the diffraction pattern in direction ϕ (Figure F.2). Notice that k_ϕ does not refer to the wavenumber of a particle, but to the wavenumber that a particle would have if its total momentum were $p_y(\phi)$.

To summarise, the steps that lead from a horizontal particle (before it enters the slit) to a unique association between a wavenumber and the direction of that particle (after it exits the slit) is summarised here (with the operation that yields each result given below each arrow):

$$\text{horiz. mom. } p \xrightarrow[\times\sin\phi]{\text{diffraction}} \text{vert. mom. } p_y(\phi) \xrightarrow[\times 1/\hbar]{\text{matter wave}} \text{wavenumber } k_\phi.$$

As we shall see, the diffraction pattern of intensity on the screen specifies how much of each wavenumber is implicit in the light at the slit. In effect, the diffraction pattern is the Fourier transform of the slit.

Diffraction as Fourier Analysis

The variable y represents position within the slit, which has a width of $d=1$ here. The slit is represented by the real function $\Psi_y(y)$, which is essentially a spatial pulse (rectangular function) such that $\Psi_y(y)=1/d=1$ if $|y|\leq d/2$ and $\Psi_y(y)=0$ if $|y|>d/2$, as shown in Figure F.3a.

Using Fourier analysis, the function $\Psi_y(y)$ can be expressed as a superposition of sinusoidal *Fourier components*. Each spatial component $e^{ik_\phi y}$ has a wavenumber k_ϕ rad/m which is uniquely associated with the direction ϕ. The amplitude and phase of the spatial component $e^{ik_\phi y}$ are specified by the value of a complex function $\hat{\Psi}_k$ at k_ϕ (as in Equation D.5):

$$\Psi_y(y) \quad = \quad C_\phi \int_{k_\phi=-\infty}^{\infty} \hat{\Psi}_k(k_\phi)\, e^{ik_\phi y}\, dk_\phi. \tag{F.5}$$

The complex function $\hat{\Psi}_k(k_\phi)$ is the *Fourier transform* of the real function $\Psi_y(y)$ (the constant C_ϕ need not concern us here, and the same applies to other constants below). In general, the Fourier transform of a real even function (such as $\Psi_y(y)$) is also real and even, so the imaginary part of $\hat{\Psi}_k(k_\phi)$ is zero here. The function $\hat{\Psi}_k(k_\phi)$ is given by the *inverse Fourier transform*,

$$\hat{\Psi}_k(k_\phi) \quad = \quad C_y \int_{y=-d/2}^{d/2} \Psi_y(y)\, e^{-ik_\phi y}\, dy. \tag{F.6}$$

Note that there is no loss of information when $\Psi_y(y)$ is expressed as in terms of its Fourier transform $\hat{\Psi}_k(k_\phi)$; the functions $\hat{\Psi}_k(k_\phi)$ and

$\Psi_y(y)$ are simply different representations of the same information. The limits of integration for the slit are $y = \pm d/2$, which yields

$$\hat{\Psi}_k(k_\phi) \quad = \quad \frac{\sin k_\phi}{k_\phi} \quad = \quad \text{sinc } k_\phi. \qquad (F.7)$$

Using $k_\phi = (p/\hbar)\sin\phi$ and setting $p/\hbar = 1$ for convenience, we have

$$\hat{\Psi}_\phi(\phi) \quad = \quad \text{sinc}(\sin\phi), \qquad (F.8)$$

as shown in Figure F.3b. Therefore, the intensity of the diffraction pattern in the direction ϕ is the *power spectrum* of the Fourier transform,

$$I(\phi) \quad = \quad |\hat{\Psi}_\phi(\phi)|^2 \quad = \quad \text{sinc}^2(\sin\phi). \qquad (F.9)$$

Crucially, this is the diffraction pattern associated with a slit.

Finally, given that $p_y = p\sin\phi$ (so $\sin\phi = p_y/p$) and setting $p = 1$, we have $\hat{\Psi}_p(p_y) = \text{sinc } p_y$. Note that ignoring constants (by setting them to 1) allows us to focus on the overall shape of each function.

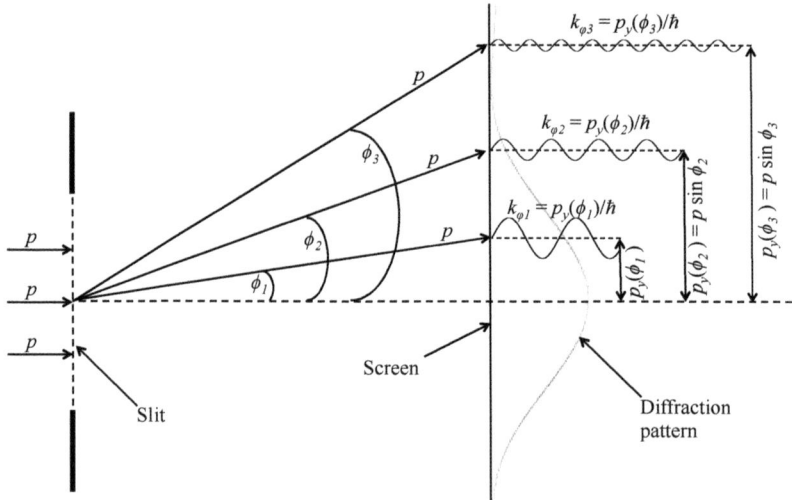

Figure F.2: Schematic depiction of Fourier optics at a slit. All particles enter the slit with momentum p and with vertical momentum $p_y = 0$. A particle that exits the slit at an angle ϕ has a vertical momentum of $p_y(\phi) \approx p\sin\phi$, so the equivalent de Broglie wavenumber along the vertical axis is $k_\phi = p_y(\phi)/\hbar$. The intensity of the diffraction pattern in direction ϕ is proportional to the number of particles that exit the slit in direction ϕ. Thus, the diffraction pattern specifies how much of each de Broglie wavenumber lies in each direction ϕ, which (for all practical purposes) is the (squared) Fourier transform of the rectangular slit.

Heisenberg's Inequality

Heisenberg's uncertainty principle is based on a fundamental result from Fourier analysis, known as Heisenberg's inequality. To understand Heisenberg's inequality, we first need to define a ubiquitous distribution: the Gaussian. The Gaussian distribution is a bell-shaped function defined by two parameters, the mean (which we assume is zero) and the *standard deviation*. If a variable y has a Gaussian distribution with mean zero and standard deviation ρ_y then the probability (density) of the value y is

$$f(y) \quad = \quad C_f \times e^{-y^2/(2\rho_y^2)}, \tag{F.10}$$

where C_f is a normalising constant (i.e. it ensures that the total probability is 1; the same applies to analogous constants below). The standard deviation measures the variability in y and determines the width of the distribution, as shown in Figure 4.7 (where standard deviation is represented by the symbol σ).

It can be shown that the Fourier transform of a normalised Gaussian function $f(y)$ with standard deviation ρ_y is another normalised Gaussian function $g(k)$ of a variable k with standard deviation ρ_k, such that

$$\rho_y \rho_k \quad \geq \quad 1, \tag{F.11}$$

with equality only if $f(y)$ (and therefore $g(k)$) is Gaussian. This is *Heisenberg's inequality*.

To place this result in the context of quantum mechanics, we can replace the functions $f(y)$ and $g(k)$ with the wavefunctions $\Psi_y(y)$ and $\hat{\Psi}_k(k_\phi)$, respectively. However, the only physical correlate of a

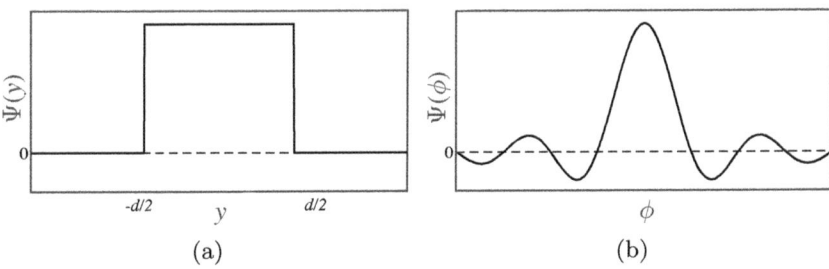

(a)   (b)

Figure F.3: Fourier transform. (a) The cross-section of a slit defines a rectangular function (Equation F.5). (b) The function $\text{sinc}(\phi) = \sin(\phi)/\phi$ is the Fourier transform (amplitude spectrum) of the rectangular function in (a); it is also the wavefunction $\hat{\Psi}_\phi(\phi)$ of light that passes through a slit (Equation F.8).

wavefunction is intensity, which is proportional to probability, which is proportional the squared modulus of the wavefunction. Accordingly, consider a slit that is transparent at the centre $y = 0$, but for which transparency, and therefore intensity, falls away as a Gaussian function of distance from the centre (Figure 4.7). Therefore, intensity within the slit is defined as the squared modulus of Equation F.10

$$|\Psi_y(y)|^2 \;=\; C_y \, e^{-y^2/\rho_y^2}, \tag{F.12}$$

which is also a Gaussian function. Suppose this Gaussian has standard deviation σ_y, so that

$$|\Psi_y(y)|^2 \;=\; C_y \, e^{-y^2/(2\sigma_y^2)}. \tag{F.13}$$

It then follows that $\rho_y^2 = 2\sigma^2$ and so $\sigma_y = \rho_y/\sqrt{2}$.

From the preceding section, the Gaussian slit provides a Fourier transform of $\Psi_y(y)$, which yields a Gaussian wavefunction $\hat{\Psi}_k(k_\phi)$ with standard deviation ρ_k, so the intensity of the diffraction pattern is

$$|\hat{\Psi}_k(k_\phi)|^2 \;=\; C_k \, e^{-k_\phi^2/(2\sigma_k^2)}, \tag{F.14}$$

where $\sigma_k = \rho_k/\sqrt{2}$ (by analogy with Equations F.12 and F.13). We can express this in terms of momentum because $p_y = \hbar k_\phi$, which implies that the spatial distribution of momentum probabilities at the screen is Gaussian with a standard deviation of

$$\sigma_p \;=\; \hbar\sigma_k \;=\; \hbar\rho_k/\sqrt{2}. \tag{F.15}$$

Therefore

$$\sigma_p\sigma_y \;=\; \hbar\rho_k/\sqrt{2} \times \rho_y/\sqrt{2}, \tag{F.16}$$

where $\rho_k\rho_y \geq 1$ (Equation F.11), which yields *Heisenberg's uncertainty principle*

$$\sigma_y\sigma_p \;\geq\; \hbar/2, \tag{F.17}$$

with equality only if $\hat{\Psi}_y$ (and therefore $\hat{\Psi}_p$) is Gaussian. Thus, a slit with a Gaussian profile of transparency produces a diffraction pattern on the screen with a Gaussian profile of intensity. As the standard deviation σ_y of the slit's Gaussian profile is reduced, the standard deviation σ_p of the screen's Gaussian diffraction pattern increases, such that $\sigma_p = \hbar/(2\sigma_y)$. It is not possible to reduce the product $\sigma_y\sigma_p$ of uncertainties in position y and momentum p_y below $\hbar/2$.

Appendix G

Wavefunctions and PDEs

The complex time-dependent Schrödinger equation (TDSE, Equation 6.39) can be expressed as two coupled real partial differential equations (PDEs) as follows. To simplify notation, we set the potential energy $V(x, t)$ to zero, which allows the TDSE to be written as

$$i\frac{\partial \hat{\Psi}}{\partial t} = \frac{-\hbar}{2M}\frac{\partial^2 \hat{\Psi}}{\partial x^2}, \tag{G.1}$$

where the x and t arguments have been omitted from $\hat{\Psi}(x, t)$. Next, we write the complex wavefunction $\hat{\Psi}$ as a sum of two terms such that all quantities in one term are real and all quantities in the other term are imaginary:

$$\hat{\Psi} = \Psi_{\mathrm{R}} + i\Psi_{\mathrm{I}}. \tag{G.2}$$

Then the temporal derivative of the wavefunction is

$$\frac{\partial \hat{\Psi}}{\partial t} = \frac{\partial \Psi_{\mathrm{R}}}{\partial t} + i\frac{\partial \Psi_{\mathrm{I}}}{\partial t}, \tag{G.3}$$

and substituting this into Equation G.1 gives

$$\begin{aligned} i\frac{\partial \hat{\Psi}}{\partial t} &= i\frac{\partial \Psi_{\mathrm{R}}}{\partial t} + i^2\frac{\partial \Psi_{\mathrm{I}}}{\partial t} \\ &= i\frac{\partial \Psi_{\mathrm{R}}}{\partial t} - \frac{\partial \Psi_{\mathrm{I}}}{\partial t}. \end{aligned} \tag{G.4}$$

Now separate the right-hand side of Equation G.1 into real and imaginary components:

$$\frac{-\hbar}{2M}\frac{\partial^2 \hat{\Psi}}{\partial x^2} = \frac{-\hbar}{2M}\frac{\partial^2 \Psi_{\mathrm{R}}}{\partial x^2} + i\frac{-\hbar}{2M}\frac{\partial^2 \Psi_{\mathrm{I}}}{\partial x^2}. \tag{G.5}$$

Equation G.1 implies that the right-hand sides of Equations G.4 and G.5 are equal,

$$-\frac{\partial \Psi_I}{\partial t} + i\frac{\partial \Psi_R}{\partial t} = \frac{-\hbar}{2M}\frac{\partial^2 \Psi_R}{\partial x^2} + i\frac{-\hbar}{2M}\frac{\partial^2 \Psi_I}{\partial x^2}, \qquad (G.6)$$

which means that the real parts on both sides are equal and the imaginary parts on both sides are equal; therefore

$$\frac{\partial \Psi_I}{\partial t} = \frac{\hbar}{2M}\frac{\partial^2 \Psi_R}{\partial x^2}, \qquad (G.7)$$

$$\frac{\partial \Psi_R}{\partial t} = \frac{-\hbar}{2M}\frac{\partial^2 \Psi_I}{\partial x^2}. \qquad (G.8)$$

Thus, as promised, Schrödinger's complex time-dependent wavefunction corresponds to two coupled real differential equations (Equations G.7 and G.8). Specifically, Equation G.7 states that the rate of change (temporal derivative) of the real function $\Psi_I(x,t)$ at position x and time t depends on the spatial curvature (second spatial derivative) of the real function $\Psi_R(x,t)$ at position x and time t, and similarly (Equation G.8) for the rate of change of $\Psi_R(x,t)$ and the spatial curvature of $\Psi_I(x,t)$.

Just as an electromagnetic wave consists of mutually dependent (physically real) electric and magnetic fields, so Schrödinger's complex wavefunction can be regarded as being composed of two (mathematically real) mutually dependent fields Ψ_I and Ψ_R. However, to observe the physical consequences of Schrödinger's complex wavefunction, it is necessary to torture it a little. For example, to find the probability that a particle is located within a small interval Δx centred on position x, we need to calculate $p(x)\Delta x$ where $p(x) = |\hat{\Psi}(x,t)|^2$.

Appendix H

Key Equations

Wave speed (ν = frequency, λ = wavelength):

$$v = \nu\lambda \text{ m/s}. \tag{H.1}$$

Angular frequency (T = oscillation period):

$$\omega = 2\pi/T \text{ rad/s}. \tag{H.2}$$

Temporal frequency (T = oscillation period):

$$\nu = 1/T \tag{H.3}$$
$$= \omega/(2\pi) \text{ cycles/s or Hz}. \tag{H.4}$$

Wavenumber (λ = wavelength):

$$k = 2\pi/\lambda \text{ rad/m}. \tag{H.5}$$

Spatial frequency is not used much in physics, but it is included here to distinguish it from the wavenumber k (λ wavelength):

$$\nu_S = 1/\lambda \tag{H.6}$$
$$= k/(2\pi) \text{ cycles/m}. \tag{H.7}$$

Planck's quanta (h = Planck's constant, ν = frequency):

$$E = h\nu \text{ J}. \tag{H.8}$$

Planck's blackbody spectrum (ν = frequency, c = speed of light, h = Planck's constant, k_B = Boltzmann's constant, T = temperature):

$$\rho(\nu) = \frac{8\pi\nu^2}{c^3}\frac{h\nu}{e^{h\nu/(k_B T)} - 1}. \tag{H.9}$$

H Key Equations

The de Broglie wavelength for a particle with momentum p (h =Planck's constant):

$$\lambda \;=\; h/p \;\text{m.} \tag{H.10}$$

Momentum of a particle with wavenumber k ($\hbar$ = reduced Planck's constant = $h/(2\pi)$, k = wavenumber = $2\pi/\lambda$):

$$p \;=\; \hbar k \;\text{kg m/s.} \tag{H.11}$$

Heisenberg's uncertainty principle states that the product of the standard deviation σ_y in position and the standard deviation σ_p in momentum satisfies ($\hbar$ = reduced Planck's constant)

$$\sigma_y \sigma_p \;\geq\; \hbar/2. \tag{H.12}$$

Complex sinusoid: the complex value of a wave at location x and time t with wavenumber k and angular frequency ω is

$$\hat{\Psi}(x,t) \;=\; e^{i(kx-\omega t)}. \tag{H.13}$$

Time-dependent Schrödinger equation (TDSE):

$$i\hbar \frac{\partial \hat{\Psi}(x,t)}{\partial t} \;=\; \frac{-\hbar^2}{2M}\frac{\partial^2 \hat{\Psi}(x,t)}{\partial x^2} + V(x,t)\hat{\Psi}(x,t). \tag{H.14}$$

Time-independent Schrödinger equation:

$$-\frac{\hbar^2}{2M}\frac{d^2 \hat{f}(x)}{dx^2} + V(x)\hat{f}(x) \;=\; E\hat{f}(x). \tag{H.15}$$

References

[1] J Al-Khalili and J McFadden. *Life on the Edge: The Coming of Age of Quantum Biology*. Penguin Random House, 2015.

[2] M Arndt, O Nairz, J Vos-Andreae, C Keller, G Van der Zouw, and A Zeilinger. Wave–particle duality of C60 molecules. *Nature*, 401(6754):680–682, 1999.

[3] W Beckner. Inequalities in Fourier analysis. *Annals of Mathematics*, 102(1):159–182, 1975.

[4] JS Bell. On the Einstein Podolsky Rosen paradox. *Physics Physique Fizika*, 1(3):195, 1964.

[5] N Bohr. On the constitution of atoms and molecules. *The London, Edinburgh, and Dublin Philosophical Magazine and Journal of Science*, 26(153):476–502, 1913.

[6] PJ Coles, J Kaniewski, and S Wehner. Equivalence of wave–particle duality to entropic uncertainty. *Nature Comm.*, 5(1):1–8, 2014.

[7] L de Broglie. Researches on the quantum theory. *Annalen der Physik*, 3:22–32, 1925.

[8] T Dimitrova and A Weis. The wave–particle duality of light: A demonstration experiment. *American J. Physics*, 76(2):137, 2008.

[9] PAM Dirac. On the theory of quantum mechanics. *Proceedings of the Royal Society of London A*, 112(762):661–677, 1926.

[10] A Einstein, B Podolsky, and N Rosen. Can quantum-mechanical descriptions of physical reality be considered complete? *Physical Review*, 47:777, 1935.

[11] R Eisberg and R Resnick. *Quantum Physics of Atoms, Molecules, Solids, Nuclei, and Particles*. Wiley, 1985.

[12] R Feynman. *The Feynman Lectures on Physics*. Basic Books, 1964. http://feynmanlectures.caltech.edu/

[13] SJ Freedman and JF Clauser. Experimental test of local hidden-variable theories. *Physical Review Letters*, 28(14):938, 1972.

[14] N Gisin. Quantum correlations in Newtonian space and time. In *Quantum Theory: A Two-Time Success Story*, pp185–203, 2014.

[15] W Heisenberg. Quantum-theoretical re-interpretation of kinematic and mechanical relations *Zeitschrift für Physik*, 33:879–893, 1925.

[16] W Heisenberg. The actual content of quantum theoretical kinematics and mechanics (English transation). *Zeitschrift für Physik*, 43:172–198, 1927.

[17] V Jacques, et al. Experimental realization of Wheeler's delayed-choice gedanken experiment. *Science*, 315(5814):966–968, 2007.

[18] MC Jain. *Quantum Mechanics: A Textbook for Undergraduates.* PHI Learning Pvt Ltd, 2017.

[19] J Kincaid, K McLelland, and M Zwolak. Measurement-induced decoherence and information in double-slit interference. *American J. Physics*, 84(7):522–530, 2016.

[20] S Kochen and EP Specker. The problem of hidden variables in quantum mechanics. *J. Math. Mech.*, 17:59–87, 1967.

[21] A Lightman. *The Discoveries: Great Breakthroughs in 20th-Century Science.* Vintage Canada, 2010.

[22] J McFadden. *Quantum Evolution.* WW Norton & Company, 2000.

[23] J McFadden and J Al-Khalili. The origins of quantum biology. *Proceedings of the Royal Society A*, 474(2220), 2018.

[24] M Minder et al. Experimental quantum key distribution beyond the repeaterless secret key capacity. *Nature Photonics*, 13(5):334–338, 2019.

[25] G Mollenstedt and H Duker. Interferenzversuch mit einem Biprisma fr Elektronenwellen. *Naturwissenschaften*, 42:41, 1955.

[26] KH Norwich. Boltzmann–Shannon entropy and the double-slit experiment. *Physica A*, 462:141–149, 2016.

[27] BY Peled, A Te'eni, D Georgiev, E Cohen, and A Carmi. Double-slit with an EPR pair. *Applied Sciences*, 10(3):792, 2020.

[28] M Planck. On an improvement of Wien's equation for the spectrum. *Verh. Deut. Phys. Ges.*, 2:202–204, 1900.

[29] D Radin, L Michel, and A Delorme. Psychophysical modulation of fringe visibility in a distant double-slit optical system. *Physics Essays*, 29(1):14–22, 2016.

[30] E Schrödinger. An undulatory theory of the mechanics of atoms and molecules. *Physical Review*, 28(6):1049, 1926.

[31] E Schrödinger. The current situation in quantum mechanics. *Natural Sciences*, 23(49):823–828, 1935.

[32] E Schrödinger. *What Is Life?* Cambridge University Press, 1944.

[33] JV Stone. *Information Theory: A Tutorial Introduction.* 2015.

[34] KC Tan and H Jeong. Entanglement as the symmetric portion of correlated coherence. *Physical Review Letters*, 121(22), 2018.

[35] GI Taylor. Interference fringes with feeble light. *Proceedings of the Cambridge Philosophical Society*, 15:114–115, 1909.

[36] A Tonomura, J Endo, T Matsuda, T Kawasaki, and H Ezawa. Demonstration of single-electron buildup of an interference pattern. *American J. Physics*, 57(2):117–120, 1989.

[37] T Young. The Bakerian Lecture. On the theory of light and colours. *Philosophical Transactions of the Royal Society of London*, 92(1):63–67, 1832.

Index

A Note from the Author

I sincerely hope that you enjoyed reading this book. If you did (or even if you didn't) then I would be grateful if you could write a review, either on Goodreads or on Amazon.

If you think you are not sufficiently expert to write a review then you are precisely the type of reader that other readers value the most. After all, a book with the words "A Tutorial Introduction" in its title should be read mainly by non-experts.

James V Stone.

Lightning Source UK Ltd.
Milton Keynes UK
UKHW021304141220
375006UK00003B/42